Sustainability in the Arctic

Proceedings from Nordic Arctic Research Forum Symposium 1993

Edited by Tom Greiffenberg

Tom Greiffenberg (ed.)

1994 Copyright by   Nordic Arctic Research Forum
                           att.: Aalborg University
                           Department of Planning and Development
                           Tom Greiffenberg
                           Fibigerstraede 11
                           DK-9220 Aalborg
                           Denmark

Cover by   Knud Thiesson, Aalborg University

Printed by   Thy Bogtryk & Offset A/S
                           Industrivej 3
                           DK-7790 Thyholm
                           Phone (45) 97871666
                           Fax (45) 97871765

Distributed by   Aalborg University Press
                           Niels Jernesvej 10
                           DK-9220 Aalborg Ö
                           Denmark
                           Phone (45) 98152928

ISBN   87-7307-471-3

Contents

Preface

This book publishes the proceedings from the second Nordic Artic Research Forum symposium. The introduction to the symposium was the issue of Sustainable Development in the Arctic. The concept of sustainable development is to be understood as in the Brundtland report, i. a. "a development which meets the needs of the present without compromising the ability of future generations to meet their own needs". This concept does not only focused on ecological crises but also stresses the importance of avoiding economic crises. The larger part of the book deals with the question of sustainability seen from different angles, but the book contains papers concerning research on other important Arctic issues as well. The book does not include all the papers presented at the symposium. A list of all presented papers can be found in appendix B. Papers from the Canadian researcher Jack Hicks and from the Danish researchers Jes Adolphsen and Tom Greiffenberg will be published separately in the research series of the Department of Development and Planning, Aalborg University together with some additional papers to get the background of the dialogue between the two parties on how to understand development planning in the Arctic.

The authors have had the possibility of revising their papers after the discussions on the symposium, but there have been no external referees and the authors themselves are responsible for what is published.

A list of the participants is included in appendix A.

The symposium and the publishing of this book was supported by the Danish Social Research Council, the Nordic Minister Council (NorFa) and Aalborg University.

The word processing assistance and the lay-out of the book was provided by Ulla Christensen and Nicolaj Pedersen, Aalborg University.

Tom Greiffenberg
Editor

Considerations on Sustainable Development in the Arctic

Lise Lyck

Abstract: The concept sustainable development is discussed and criticized. Main problems relating to the concept are outlined. Among the problems intergenerational equity is treated theoretically in a legal perspective and its operationalization is dealt with. The needed change in decision making towards greater emphasis on long term perspectives is discussed and the possibility of using environmental impact assessments as a means to achieve this is outlined. Furthermore, the character of problems of present scientific knowledge is treated and the problems related to internalization of externalities and to less individual economic space are dealt with. Finally, sustainable development in the Arctic is set in focus. The meaning of Arctic is outlined and the concepts of "Arctic economy" and "Arctic economies" is defined. The main impacts the Arctic World meets are presented and analyzed in a sustainable development context. It is concluded that the Federation of Russia can develop an energy supply to the EEC similar to the Canada/US relations. It will increase the role of mega projects in the Arctic, but the experiences, especially from Prudhoe Bay, will increase cooperation between the largest corporations and the most powerful nations, reducing competition and giving room for sustainable development with respect to investments within a framework of a balanced economic growth. Environmental standards and safeguards on Arctic nature and species will improve but Arctic econmies will follow the international market economy and sustainable development will be subordinate to the market economy.

1. Sustainable Development

Overview 1: Sustainable Development

Sustainable development:		Development that meets the needs of the present without compromising the ability of future generations to meet their own needs.
Interpretations :	1	Balanced economic growth (Our Common future) (The 1969 Stockholm Conference)
	2	Ecological, giving up economic growth as a rationale
	3	Human ecological (allow for and facilitate spiritual, social and cultural development) (ICC 1992)
	4	Linking of tenure with stewardship (a way of life in which humans are seen as constituents of natural systems rather than humans ruling over them) (Griffith, Oran Young)
	5	Eco-feminists (Linking domination over nature to domination by men over women)

The World Commission on Environment and Development chaired by the then Norwegian Prime Minister Brundtland focused on the concept "sustainable development" in Our Common Future and defined it as "*a development which meets the needs of the present without compromising the ability of future generations to meet their own needs*".

Popularly, sustainable development has since often been understood as a development that avoids ecological crises, but it is important to be aware that this is a much more limited concept than the Brundtland report concept which includes development that avoids both economic and ecological crises, i.e. the Brundtland report concept *does not imply a zero economic growth but a balanced economic growth.* In this way it corresponds to the understanding of development already presented in the 1969 United Nations Stockholm Conference on Social Policy and Planning in National Development. Discussions on the meaning and implementation of the concept have taken place for more than 20 years though the debate has been intensified after the publication of the Brundtland report in 1987.

In the later years the rationale underlying sustainable development as the perceived need to balance economic growth with protection of the environment has been sharply criticized for saying nothing about the net costs due to degradation of certain elements of society and its natural resources. The criticizers advocate that environmental sustainability requires recognition of giving up economic growth as a rationale. Inuit Circumpolar Conference in their 1992 "Principles and Elements for a Comprehensive Arctic Policy states "When undertaking developmental or other activities of any nature, planners and decisionmakers must not simply view the Arctic as an exploitable frontier ... Northern development must refer to more than economic growth. It must allow for and *facilitate spiritual, social and cultural development".* It stresses the human ecological perspective to environmental issues. The interpretation of sustainable development as" ... a way of life in which humans are seen as constituents of natural systems rather than masters ruling over them...." is found among political scientists like Oran Young and Frank Griffith (Arctic, March 1993). Economists go further, linking domination over nature to domination by men over women.

To summarize, the concept sustainable development is a still more ambiguous and philosophical concept of *normative character, not a technical standard.*

In this sense it is similar to and can be compared to the use of the concept "subsidiarity" in the EC. Though subsidiarity is an ambiguous philosophical concept with roots both in the Bismarck Preussen and in the Catholic Church since the 1930s it is found in the Maastricht Treaty without any further interpretation.

Concepts like those raise a lot of problems: Can they be defined in an acceptable way and can they be made operational in a social context?

Even when sustainable development is perceived as balanced economic growth severe problems to deal with persist.

Among the problems are 1) *intergenerational equity*, 2) *decision making must be changed* for greater emphasis on the long-term viability of goals and objectives, 3) *too little scientific knowledge* of deviations from what is "normal" and 4) internalization of *externalities* and less individual economic space.

All problems end up in what is concluded in the final analysis of the Brundtland Report that "sustainable development must rest on political will", (note 1 at 9).

In the Western World many people have been dissatisfied with the governments' environmental policies and have tried through grassroot movements and the establishing of new political parties to give the environment a higher priority. The grassroot movements have been more successful than the new environment political parties in attaining political influence. People's criticism emphasizes that 1) existing standards for environment protection are too permissive, 2) sanctions for non-compliance with standards are too mild or not enforced, 3) all economic activities which can harm the environment are considered innocent until proven guilty and by that time often irreversible, 4) many compounds are permitted without rigorous assessment of risk, and 5) governments are too tolerant in not reducing known hazards related to production. In *economic terms* you could say that the environment in the Western World is a kind of "opposite merit want", which the politicians are slow to recognize.

In the rest of the world people often find that the governments should concentrate on efforts to increase economic growth and consider the environment as a sort of luxury. The governments are often more environmentally oriented than the population which means that environmental protection and preservation can be considered a merit want.

Beside the mentioned problems, the concept of sustainable development has also been questioned in a more fundamental way. Adrian Tanner e.g. argues that "the Brundtland Report offers no conception of development as a social process within which cultural ideas or values have a place" and "that the situation though described in terms of *rights* and *socially desirable* are otherwise treated in universal and culture-free terms". To exemplify this, Adrian Tanner mentions "the population growth which calls for campaigns to strengthen the social, cultural and economic motivation for couples to have small families" (Our Common Future, note 1 at 67-68). He holds "that development-oriented economies have cultural values and institutional forms that are inseparable from patterns of boom and burst and firm ecological crisis; these phenomena are not accidents, but are generated by the development process itself". In this perception he is totally in line with anthropology, ethnography and part of political science and economics explaining that modernization (economic development) though its origins and process may make it a unitary, worldwide, multinational phenomenon, has a specific historical context and a specific set of economic and cultural relationships which implies that development problems can only be solved, if at all, through understanding of these and of how this process is bound to create new cultural situations out of the previous ones. As a consequence sustainable development at the general level must be connected to human rights, economic rights and to self-government and at the specific level to the single cultures and peoples.

The concept of sustainable development is also narrowly connected to economic theory and economic history. The correspondence back in the late 1700 between Thomas Malthus and David Ricardo pose the question: can

11

economic growth continue without regulation or, put in another way, can technology compensate for lack of resources. They answer it in opposite ways, Malthus calling for regulations and Ricardo believing in the free initiative. The same discussion was the theme for the Club of Rome in the 1960's in "Limits to Growth". It can also be argued a similar discussion recurs when sustainable development is compared to free market economists such as e.g. Coase. This old and ongoing discussion is one of the reasons for the ambivalent attitude to sustainable development in economics and political science, and also in practice.

2. The problems in making sustainable development in an operational concept

The problems in making sustainable development an operational concept are as mentioned earlier mainly related to 1) intergenerational equity, 2) change of decision making for greater emphasis on long term considerations and less weight on short term profits, 3) the present level of scientific knowledge and 4) interlization of externalities and less individual economic space.

Intergenerational equity is a core element of sustainable development and Article 2, entitled "Conservation for Present and Future generations" provides that "States shall ensure that the environment and natural resources are conserved and used for the benefit of present and future generations" (Environmental Protectionism and Sustainable Development, World Commision on Environment, note 18 to 44).

The intergenerational equity principle is closely connected to the *non-discrimination* principle but it is a more comprehensive principle, as it both calls for equal treatment of everyone but also sets an instrumental standard by directly prescribing that subsequent generations shall be under no greater disability than the present ones.

12

Intergenerational equity has more aspects. It includes that subsequent generations should have as broad a range of options available as the present generation. It also requires that the options of future generations should be at least as real as the present options. Furthermore, it requires that cost and benefit from projects and policies are intergenerationally neutral.

It can be argued that intergenerational equity can to be found in the general principles of utility regulations as e.g. in the argument for a cost benefit balance of the balance of payments and for balance of public expenditures and revenues. Weiss (1984) has argued for a trust approach, understanding intergenerational equity as imposing a more or less formal trust on the present generation. The principle can also be found as the underlying principle of many fisheries agreements and of international agreements dealing with the harvesting of particular species or stocks. It can be specified as sustainable yield in the form of optimum sustainable yield, maximum sustainable yield (maximizing the physical measure of the catch), as maximum sustainable effort under circumstances of no external rent or external subsidy (no control of access to the stock) or maximum sustainable effort with both internal and external subsidy (maximum sustainable physical effort is expanded to catch a small stock of fish).

To summarize, the intergenerational principle is a powerful means which is and can be implemented in both national and international laws and economy.

Change of decision making for greater emphasis on long term perspectives and less weight on short run profits

Such a change demands a change of economic attitudes and behaviours away from materialism. Due to competition it has to be a massive change if it is to be viable. It is out of range to attain such a change unless it is supported by public initiatives in price-setting and taxation. Due to this need for public regulation and policy changes are unlikely to penetrate the economy towards the year 2000 as the international trend is in favour of more market based approaches. This means that this change will not occur among economic agents. On the contrary, it must be expected to occur in relation to investments by persons or associations actively pressing for a higher environmental standard. As a consequence the usual investment criteria will be replaced or supplemented with other types of information. In this development Environmental Impact Assessment as a tool will often be used.

An Environmental Impact Assessment is a pragmatic way and method to implement the basic principles in the 1972 Stockholm Declaration on Human Environment in form of the rule of good neighbours (droit de voisinage) and the maxim "sin utere ut alienum non laedes".

Environmental Impact Assessments are used to identify environmental impacts so that the decisions can be taken in the knowledge of environmental consequences.

Environmental Impact Assessments is a means of preventing environmental problems at the planning stage of an activity, hence it can only be applied to new projects which has given raise to criticism of the Environmental Impact Assessment as instrument. There are still many problems related to Environmental Impact Assessments, especially in terms of lack of 1) authority and accountability, 2) adequacy of scientific studies, 3) public access to the elaboration of environmental impact assessments and 4) how much the environmental impact assessment affects the investment decision.

The change of decision making also requires new information for decision making at the general level for which reason a still stronger pressure for green national accounts will be seen.

Present scientific knowledge

The problems are closely connected to the lack of knowledge of baseline conditions and trends and deviations. This in turn is related to the lack of long-term monitoring and relevant data. Some of the problems, especially in the Arctic, could be reduced if more military knowledge was given free to civilian purposes. In spite of the East-West detente and the collapse of the

Soviet Union as a superpower, most military knowledge and monitoring data are still classified.

In the long term, more knowledge will be available due to comprehensive monitoring programmes recently initiated like e.g. AMAP and due to many more resources dedicated to environmental research.

Internalization of externalities and less individual economic space

Internalization of externalities raises both national and international problems. At the national level the problems are closely related to the distribution of rights and incomes and to the political power of the different social groups of the society. At the international level it is dependent on the economic force and the political will of the states. It is visualized in the difficulties of making binding international agreements on environment and, looking for instance at the Rovanieni process it is easily seen that it is utmost problematic. Furthermore, it is evident that a sustainable development implies less individual economic space which is totally against the idea of the market economic approach which is gaining ground. A tendency has developed that the old distinction between workers and capitalists has been replaced by a distinction between environmentalists and capitalists.

3. Sustainable development in the Arctic

Overview 2: Structural framework

The Arctic includes arctic economies and arctic territories.

Arctic economies are Arctic societies in which a modernization has taken place and with autonomy in decision making. They can have a *unitary structure* with common identity and values :

> Iceland
> Greenland
> The Faroe Islands
> Nunavut

or have a more *dual structure*, including indigeneous peoples and other groups without common identity and common values :

> Alaska
> Canadian provinces

Arctic territories are mainly part of their respective nations, i.e. common identity and common values with the nation

> North of Norway, Sweden and Finland
> the Northern part of the Federation of Russia

Arctic territories are now in the phase of developing and expressing own values and identities which means that Arctic territories move in direction of being Arctic economies.

Arctic economy can be defined as a divided economy based on exchange of resources for other goods, including both *traditional economy* with exchange having a subordinate character and with dependency on transfer income from the State and an *advanced high technology production* with price setting and investment decisions mainly decided outside the Arctic.

Initially, it is necessary for a relevant discussion of sustainable development in the Arctic to be aware of the structure of Arctic economy and Arctic economies and in this understand what is meant by Arctic.

To start with the latter, Arctic in this context is not a precisely defined territory, as its southern boundaries are related to life conditions and socio-economic conditions. Since the Cold War, the State, especially the superpowers, focused on control of territories and sovereignty which implied a heavy military presence in the Arctic and also widened the Arctic area in southern direction. Furthermore, the growth of the public sector, especially from the end of the 1950's, has been connected with the state having an increased redistributive function which has implied socio-economic claims in still more southern areas in order to be considered northern and to obtain increased public transfer incomes. As a consequence, territories considered Arctic/Northern have never been larger than now.

Arctic economy can be defined as a sort of divided economy based on exchange of resources for other goods, including both traditional economy with the exchange having a subordinate character and with dependency on transfer income from the State, and of an advanced high technology production with price-setting and investment decisions mainly decided outside the Arctic (Lyck 1991). Today this implies that Arctic economy includes mainly mega projects on the one hand and a sort of traditional economy on the other hand.

Arctic economies are societies in the Arctic in which a modernization has taken place and with autonomy in economy decision making. Such economies are few, including Iceland, Greenland, the Faroe Islands with a unitary structure, and Alaska and most Canadian provinces with a dual structure, i.e. including indigeneous peoples and other inhabitants belonging to distinct societies without common identity or common values or political priorities. When the new Nunavut will be a reality it will also be an arctic economy with a unitary structure.

The *Arctic territories* in the Federation of Russia and in the North of Norway, Sweden and Finland are mainly part of their respective nations and their economy is to be studied within a regional economic framework.

The whole Arctic world meets three main impacts from outside: 1) The Arctic world is a net receiver of pollution from outside due to still more extensive activities south of the Arctic. As nature and life conditions are much more sensible in the Arctic, this generates special and severe environmental problems. 2) The Arctic societies are still remote and exotic and in this way victims for opinions and attitudes formed and formalized south of the Arctic. This means that development of traditional economy and harvesting, trapping and catching methods in a mainly traditional economy is extremely vulnerable to "opinions and attitudes" which originate south of the Arctic. 3) Market economic development. The market economy being introduced in the former Soviet Union and Eastern Europe implies a heavy increase of market considerations and weak state regulation. It is to be realized that the Federation of Russia of which two thirds of its territory is in the Arctic can develop an energy and resource delivery function to the "rich" EEC of the same character as that of Canada to the US. This will give a strong pressure for more mega-projects. The market development will also decrease public transfer income to the Arctic and thereby diminish the traditional economy. The mega-projects are so enormous that a close cooperation among enterprises and states will be needed. Initiating these projects will require that the land claim questions in the Arctic are solved. It is likely that mega-projects, especially those related to energy, will continue to give the Arctic an important role in the world economy.

What position does sustainable development hold in this development in the Arctic?

For *mega-projects* the experiences from Prudhoe Bay concerning cost overruns (900 percent), financing problems, technical problems, establishing of consortiums, state-cooperation and, finally, the Exxon Valdez oil spill catastrophe in Prince William Sound give unambiguous, strong signals: that cooperation is a must and that environmental considerations must have a high priority. In other words in the long run it calls for international agreements and environmental standards which can be in line with sustainable development goals in respect of securing a balanced economic growth and stability in the energy delivery systems. The central actors having economic interests in this development will be the oil corporations and Japan, US, Canada, EC, the Federation of Russia and the arctic economies.

The new Round Table process in Canada, in which government, industry, and environmental interests at the highest levels and in all jurisdictions are brought to the table with a view to defining a joint approach to the challenge of sustainable development can be seen as a first step in that direction. "Although the results are thus far inclusive, and quite uneven from one jurisdiction to the next, there are indications that the progress may hold important lessons for the United States and could ultimately affect the relationship of the two countries" (1991 North American Environmental Assessment Group Meeting: Conclusions from the Chair).

Overview 3 : Sustainable development in the Arctic ?

Arctic economy :   More mega-projects less traditional economy

   →   In the Federation of Russia market economic development with environment having a low priority.

     In the rest of the Arctic - based on the Prudhoe Bay experiences - a sustainable development with a balanced economic growth.

Arctic economies :   More market economy oriented, trying to demonstrate better economic performances.

   →   Sustainable development subordinate to economic growth.

Concerning *Arctic economy* the establishment of new mega-projects is likely to increase cooperation between the largest corporations and the most powerful nations, reducing competition and giving room for sustainable respects to environment within a framework of a balanced economic growth in the Arctic except for the case of the Federation of Russia where the corporations can continue with mega-projects without paying much attention to sustain-able development. (Vartanov distinquishes between official priorities and factual priorities ranking environment with a high official priority and a low factual priority).

For *Arctic economies* the signals are more mixed. For traditional economy sustainable development considerations seem to be taken into accounts when agreements on preservation of nature and species are made, but for the general economic development the market economy principle is followed. It is the case for Greenland where a new free market oriented policy took over at the end of 1991. This is in fact also what was recommended for Alaska, facing less oil money, by Scott Goldsmith in the summer 1992, calling for safe landing: A fiscal strategy for the 1990s involving cut spending, use of permanent fund earnings, encouragement of economic development, levying of taxes and conservat ion of and investment in windfalls.

To conclude, for Arctic economies the tendency seems to be to follow the international market economy trend and to subordinate sustainable development to the market economy. This choice is not surprising considering the Arctic economies' needs and obligations to show better economic performance being autonomeous.

References

Beanlands G.E. & Duinker P.N., 1983, *An Ecological Framework for Environmental Impact Assessment in Canada*, Federal Environmental Assessment Review Office.

Bonbright J.C., 1961, *Principles of Public Utility Rates*, New York, Columbia Press.

Cassils, J. Anthony, 1990, Structuring the Tax System for Sustainable Development in *The Legal Challenge of Sustainable Development* (ed) Saunders, Calgary.

Chance, Norman A., 1993, Sustainable Utilization of the Arctic's Natural Resources, *Arctic*, Volume 46, No 1, March 1993.

Davidson, A. & M. Daence (eds), 1968, *The Brundtland Challenge and the Cost of Inaction*, Halifax: Initiative for Research in Public Policy, 1968.

Goldsmith, Scott, 1992, Safe Landing: A Fiscal Strategy for the 1990s, *ISER Fiscal Policy Papers*, No 7, July 1992, University of Anchorage.

Hennesy, J.L. and Moltke, K., *1991 North American Environmental Accessment Group Meeting*: Conclusions from the Chair.

Hirsch, F., 1976, *The Social Limits to Growth*, Cambridge, Harvard University Press.

Inuit Circumpolar Conference, 1986, *Towards An Inuit Regional Conservation Strategy*, Kotzebue, Alaska.

Inuit Circumpolar Conference, 1992, *Principles and Elements for a Comprehensive Arctic Policy*.

Lyck, Lise, 1990, *Economic Development and Circumpolar People*, 7th Inuit Studies Conference, University of Alaska, Fairbanks, August 1990.

Lyck, Lise, 1990, *Prospects for Sustainable Development in the Arctic in the Light of International Political and Economic Changes*, The 3rd Northern Regions Conference, Anchorage, September 1990.

Lyck, Lise, 1990, *Canada's Position and Rôle in the Circumpolar Development*, International Canadian Studies Conference, Kingston.

Lyck, Lise, 1992, Perspectives on Arctic Economy and Arctic Economies towards year 2000 in *Nordic Arctic Research on Contemporary Arctic Problems* (ed) Lise Lyck, Aalborg University Press, ISBN 87-7307-452-7.

Lyck, Lise, 1992, *Economic and Arctic Development in Greenland since Introduction of Home Rule in Greenland 1975*, 8th Inuit Studies Conference at Laval University, Quebec, October 25-28, 1992.

Lyck, Lise, 1992, *Greenland Economic Development seen in an Arctic and in an International Perspective*, The first International Arctic Social Science Association, Laval University, Quebec, Canada, October 28 - November 1, 1992.

Pepper D., 1984, *The Roots of Modern Environmentalism*, London, Croom Helm.

Meadows D. et al., 1972, The Limits to Growth, New York, Universe Books.

Nord, Douglas C., 1990, Creating Political Institutions for the Periphery, *Circumpolar Perspectives*, Western Regional Science Association, Hawaii.

Rees, W.E., 1988, *A Pole for Environmental Assessment in Achieving Sustainable Development*, 8 Env. Imp. Ass. Rev. 273.

Sagoff, M., 1988, *The Economy of the Earth: Philosophy, Law and Environment*, Cambridge, Cambridge University Press.

Swaigan J. (ed), 1981, *Environmental Rights in Canada*, Toronto, Butterworths.

Tanner, Adrian, 1990, Northern Indigenous Cultures in the Face of Development, in *The Legal Challenge of Sustainable Development* (ed.) Saunders, Calgary.

The World Commission on Environment and Development, 1987, *Our Common Future*, Oxford, Oxford University Press, (Chair: G.H. Brundtland).

Weiss, E.B., 1984, *The Planetary Trust: Conservation and Intergenerational Equity*.

Vartanov,R, 1992, The Arctic: A new Russian Policy Emerging, presented at *the First International Arctic Social Science Association Conference*, Laval University, Quebec, Canada October 28-31, 1992.

Implementation of Sustainable Development - Methodological and Conceptual Considerations Concerning the Measuring of Sustainability

Rasmus Ole Rasmussen

Introduction - purpose and starting point

Sustainable development has in many ways developed into a current phenomenon of latest fashion. This is on the other hand not the worst fashion to follow, as the environmental problems and the need of sustainability demands all the attention it can get. But a general problem in this connection is the low level of concreteness it often is used in.

Sustainable development as a philosophical or moral principle is undoubtly a fine thing, and as a superior guide for the decision process etc. it can contribute to a positive rethinking of the development process, putting this into a frame of responsibility towards the resource situation. But an unfolding of the concept into practical guidelines calls for much more concrete conceptual considerations as well as a functional framework.

In short: The principle of sustainable development should be measurable; it should be possible to decide whether or not a given development is sustainable, and to what extent a certain development results in a more or less degree of sustainability.

It is the purpose of this contribution to make an introduction to a discussion about some of the conditions for *sustainable development* especially concerning the possibility of *measuring* the sustainability. This is done through four different approaches:

Firstly there will be a short introduction to several ways of dealing with natural conditions as an integral part of development. In many of today's discussions about sustainable development it looks like this way of thinking is an invention of the Brundtland report. This is certainly not the case.

Secondly there will be some considerations about how to measure/register sustainability. It is one thing to have sustainability as a principle and to use this as a moral guideline for the society, but something quite different to develop a concept which is operationable. One of the considerations is the perspectives of letting sustainability be an integral part of a "report book" or measurement of development defined by the national product.

Thirdly there will be an introduction to a few examples of how an extensive registration and analysis of natural conditions for, and natural consequences of, development to some extent intends to be an operationalization of the concept of sustainability. These examples are from the North Atlantic region.

Fourthly there will be an outline of a strategy for the establishing of a "green" national product for Greenland starting with a discussion of potential sources of information and some considerations about the quality of information in the context of a green national product. Thereafter there will be an encircling of a few key concepts which are considered of decisive importance for as well the development in Greenland as well as for the Greenlandic environment and natural resources situation.

When reading this contribution it is important to keep in mind that it is not a "final" report with definite conclusions. The primary purpose is to establish a starting point for a discussion of *concepts*, *aims* and *means* for the future work.

The principle of sustainable development

Many of the present discussions about sustainable development have their originate in the definition of sustainability formulated in the Brundtland report.

It is obvious that the Brundtland report has had an immense influence on the discussion of sustainability. On the other hand, the Brundtland report has not invented this way of seeing things.

Some of the crucial elements of the idea of sustainable development have been central points of interest for many scientific activities both before and certainly to a great extent after the famous *A Blueprint of Survival*[1] which in 1972 was, if not the starting point, then at least an essential catalyst for the environmental way of thinking and for the active participation in environmental discussions. At the same time the appearance of this report was expressive of the mainstream development which emerged during the second half of the sixties.

The following should not be considered as "reviews" of the approaches mentioned as only a few representative examples from the manifold literature will be mentioned and discussed very briefly. It has much more the purpose of broadening the view of how to approach the question measuring sustainability.

Limits to Growth - The development pessimists

A Blueprint of Survival[2] raised the question of the limitations of the nature to the continuation of a development based on an expanding interaction with nature. It also made suggestions for alternative conditions for the development.

With the Club of Rome and their Limits to Growth[3] this point of view was operationalized. Based on the actual pattern of the exploitation, of natural resources a projection of a continuous development of this predicted that several resources would be used up before the turn of the century unless a substantial change in the resource exploitation was introduced, even with the assumption that the development of new technology would expand the reserves several times compared to the ones known at that time.

Seen with the knowledge we have today about the resource situation and some of the dynamics of the technological changes, for instance the marked changes in technologies and material base for industries which at that time were responsible for the major exploitation of such things as copper, silver, etc., it is obvious why some of the predictions did not come true.

But already when they were published a lot of discussions was going on. Generally the critique was about the mechanical forecasting of development and thereby the severe limitations of the results. Both were focusing on the limitation of the resource and did only to a lesser degree discuss the conditions of the growing amount of waste and pollution.

This was, on the other hand, in the focus of a number of "grass-root" organizations such as the English *Friends of the Earth* and the Danish *NOAH*. The latter stepped into the arena with the book "Nogle oplysninger om den jord vi sammen bebor"[4] in which the ecological approach was introduced. They saw the dynamics of nature as intermingled with the dynamics of development of the society. But nevertheless: the idea of the earth as a spaceship in which the society's metabolism with nature has to develop, is central to today's discussion of sustainable development.

Many of the fundamental ideas from *Limits to Growth* are essential in today's discussion about sustainability, and often with arguments which in many ways resemble the ones dominating back in the 70s[5] ending up in making sustainability a question of morality. On the other hand there is also a development in arguments, especially qualitatively. Many new elements are present in today's concern about the limited resources; what the present discussion to some degree lacks is the quantitative approach.[6]

The principal resemblance between then and now is the basic moral approach to the environmental problems and the limitation of the resources, and thereby underlining the need of a general understanding of the dynamics of the nature and of the society's activities in nature. And at the same time also a certain degree of anxiety of development.

The principal difference is the lack of operationality; where *Limits to Growth* had its starting point in quantitative considerations and used theses as a basis for operational discussions, to some degree lacking important qualitative elements, the situation is almost the opposite with the discussions about sustainability today.

The concept of Carrying Capacity

The concept of *Carrying Capacity* has been, and still is, a traditional theme in anthropology, human ecology, geography and related scientific fields, dealing with traditional cultures and their dependence of / interrelations with the natural surroundings.

Broadly speaking the *Carrying Capacity* can be characterized as the potential production in a (terrestrial) ecosystem without introducing ecosystem-external matter and energy, but with the possibility of man influencing the ecosystem in different ways.[7]

When comparing the ideas behind the concept of *Carrying Capacity* and the principles of sustainable development at least one basic difference becomes obvious, and that is the lacking in the discussions of sustainable development of some of the dynamic elements found in the concept of carrying capacity.

A great quality of the *Carrying Capacity* compared to both the Limits to Growth-approach and the Sustainable Development is the tradition of a multi-dimensional analysis covering quantitative, qualitative as well as cultural aspects of the development and the relations to nature.

The limitations of the traditional way of using the concept shows in three ways: when it comes to the introduction of ecosystem-external sources of matter and energy, when it comes to the dynamics of the technological development, and when it comes to the question of evaluation of man's influence on the natural system.

Today's discussions of sustainability mainly deal with the limitations of the natural conditions, and the organizational consequences of this. Only very seldom the discussions approach the limitations set by culture and traditions in the man-nature relations. The inspiration of the traditions from the Carrying Capacity-approach could be the introduction of these elements to the discussion.

The concept of potential production - carrying capacity in a technological framework

In relation to research in the geographical aspects of a human ecological approach there has been a general discussion about the *concept of potential production* which means an understanding of a man/society/culture/nature relation where the development of technology represents the key to the understanding of the dynamics. Broadly speaking this means that the limits to growth are seen as a mixture of absolute limitations set by nature and technology and relative limitations set by society and culture.

Two approaches to the concept have been introduced.

The first approach is closely connected to the applied landscape research and landscape ecology[8], and resembles in many ways the principles of entropy in physics.

In this approach the general potential P of a landscape - the framework in which human activities are taking place - could be determined by the equation:

$$P = R + G + B + K$$

where

R = Energy received from radiation
G = Landscape energy / Gravitational energy
B = Energy in chemical, physical and biological processes
K = Energy due to human activities

When the potential of the landscape is to be released, the amount of necessary human work could be expressed as:

$$P = T + U + F + D + N$$

where

T = Technical measures which enable exploitation to take place,
U = Costs in processes needed in order to prepare material for the market's requirements,
F = Secondary installations such as homes, institutions, etc.
D = Destructive differentiation due to the jeopardizing utilization for other purposes, and
N = Economic profit/loss of the process.

The second approach[9] to the concept of potential production puts the focus on technology and technological development to a much greater extent. It does not see the potential as a fixed amount of energy or matter, but as an amount of utility value extractable from nature by means of human labour and limited by nature and technology.

A *single potential production* is in this terminology the possible production defined by the limitations posed by nature with a given technology, with technology defined in a broad sense covering both the technical and the organizational aspects. With the development of technology the potential production expands.

31

The *complex potential production* is the interwoven pattern of possible productions defined by the situation where the products, byproducts or waste from one production at the same time is the base for another production.

An advantage of the concept of potential production is obviously the interconnection of qualitative and quantitative elements as well as an understanding of a dynamic development within these categories.

A disadvantage of the concept is the difficulty of introducing the more long-term consequences of a given production, both regarding the resources and the waste management due to the problems of making forecasts of the technological development which is necessary for the evaluation of the consequences.

The potential production approach as described above has been developed into a methodology called *The Human Ecological description method*[10]. According to this, the natural conditions and limitations for production and reproduction are determined by the interrelations between nature, technology and society.

The strength of this approach is undoubtedly the methodology of analysis which combines the dynamics of nature and society by means of technology. This, on the other hand, is also its weakness because of the problems of getting the knowledge of all the single dynamic elements necessary for the full description.

The Soviet experiences - the development optimism

In spite of the general opinion that discussions about resources and environment are primary a phenomenon known in the West, the case is actually quite different. Also Eastern Europe and the Soviet Union have had their share. Not in the same way as in the West where we have seen marked public discussions, "grass-root" participation and obvious antagonism between different interests. In Eastern Europe and the Soviet Union the "public" participation has to a great extent been within officially supported and recognized organizations, and with marked political limitations.

But nevertheless - there have been marked discussions not only with sporadic appearance but rather systematic, starting in the late 50s and a marked development during the 70s.

At the beginning it was concentrated on the intensification and more rational exploitation of limited resources, but during the 70s it covered a broad spectrum of discussions with a focus on a more holistic approach to the cycles of matter and energy, and a transformation of this approach to measurable and comparable units. It is discussions which in many ways have been absent - and missing - in the West. In the East these discussions have been an integral part of the need for quantitative measures incorporable in the systems of planning[11].

The planned economy has demanded the creation of functional elements which have forced both development of methodology and concepts which were operational. This development did not necessarily turn out to be helpful in solving some of the environmental problems, but this is a discussion which will be introduced later. But it poses an important question concerning sustainability as a principle: The creation of a methodology and the necessary concepts are only the technical means in the development process; only when each individual acts responsible both in personal and in environmental issues, the sustainability will be a potential reality.

The discussions in GDR - Environmental technocracy

No matter how paradoxical it may sound, some of the most consistent works concerning implementation of sustainability are to be found in the former GDR.

In spite of the documented massive environmental destructions and waste of resources especially characteristic of the development during the last 20 years of its existence, and still more obvious because of the reduction of the environmental problems in Western Europe due to a massive public activity.

During the 70s and 80s a massive activity was concentrated in research in economics and geography concerning the transformation of complex information about environment and resources into first of all economic categories applicable to the planning system[12].

Accounting of matter

One way of keeping track with the development in environment and resources has been through a system of *accounting of matter* in which every level in the production (and reproduction) process has to account for all elements, chemical compound and energy going in and out of the system. An illustration could be the discussions in *Zur Materialökonomie*[13] where the present legislation concerning this approach which has been implemented is described. Especially due to the discussions about the need for an intensification of the use of the scarce domestic resources at the beginning of the 80s there were several attempts of a further development of both the theoretical and the technical foundation of this method.

The obvious advantage of this method is the possibility of an exact tracing of all elements, and thereby the creation of a management tool which at the same time enables accounting of matter and energy, the optimization of processes, minimization of waste production, etc. The present discussions in Denmark about taxation of CO_2 is only a vague shadow of the massive discussions in GDR at the beginning of the 80s, but a good illustration of the potentials of the method.

One of the more marked disadvantages of this method is that it does not open up an evaluation of alternative resource bases for a given production of an article for use. The method offers the possibility of evaluating different articles one by one, but not of comparing them. As a planning instrument for the bureaucracy it is usable, but useless as a more dynamic instrument for the development of alternatives.

34

Economic categorization

Another approach is represented by the works of *Hans Roos*[14]. In these the circulation of matter and energy are transferred into economic categories based on the amount of human labour in connection with the complete metabolism of man with nature as specified in these 7 central points:

1. The search for and scientific development of resources
2. Implication of stocks and other natural resources into the reproduction of the society
   a) Technological preparation for exploitation
   b) Infrastructural preparation for exploitation
3. Production
   a) Extraction and first preparation - primary resources
   b) Transformation of matter and energy
   c) Shaping
   d) Compounds of transformed and shaped materials
4. Consumption
5. Recycling of waste material and energy from production and consumption
   a) Ecological re-circulation
   b) Re-circulation governed by society
6. Productive re-circulation of waste material and energy - secondary resources
7. Regeneration of resources - material and energy
   a) Natural regeneration of resources
   b) Regeneration of resources governed by society

The advantage of this approach is that beyond a valuation of the circulation of matter and energy, it also contains the dynamics of society, i.e. technological changes, research, development, etc. which creates the alternatives to the present dependence of resources, as well as the costs of regeneration of both resources and environment. And in this way it comes rather close to the crucial points emphasized in connection with discussions about sustainable development.

Summing up

The experiences from the examples from the Soviet Union and GDR can be summarized in two main points:

Firstly they demonstrate that there are a number of possibilities in creating operational parameters which enable a continuous monitoring of the development of the resources and the environment.

Secondly history shows that it is not enough to establish these operational parameters. It takes much more to create sustainability - this is evident when looking at the present development of environmental problems in Soviet Union and GDR.

It is possible to blur, redefine, evade, etc. any operational parameter in order not to observe the intentions, no matter whether it is in relation to environmental relations or not. There need to be an engagement and a concern for the proper use.

The Brundtland parameters

In order to get around this discussion the Brundtland definition of sustainable development should be shortly presented.

According to the Brundtland report sustainable development: *"is development that meets the needs of the present without compromising the ability of future generations to meet their own needs"*[15].

What this implies is described by the seven principles below:

1)   A political system which secures the effective participation of the citizens in the decision process,

2)   An economic system which is able to produce surplus and technological knowledge in a selfcontained and sustainable way,

3)   A social system which guarantees solutions to the problems stem-
      ming from an inharmonic development,

4)   A system of production which respects the obligations to retain the
      ecological base of the development,

5)   A technological systems which evolves or searches for new
      solutions,

6)   An international system which supports sustainable patterns of
      trade and financing, and

7)   An administrative system which is flexible and with the ability to
      response ...

The quality of this approach is obviously the extent of morality contained
within the conceptual frame; in many ways it can be said that the implementation of sustainable development calls for a triple revolution: an economic,
a social and a cultural change of approach to nature.

The problem is the lack of operationality: none of the above-mentioned
specifications points to any solutions or just indications of how to get on;
How do you decide whether a certain development contributes to a sustainable development or not ? How do you tell whether one development is to
prefer compared to another ?

Conclusions

The ideas behind the principles of sustainable development are certainly not
new!

During the last 20-30 years several independent research themes and
research traditions have emerged, each of them stressing some of the crucial
elements of sustainability and thereby with a possibility of contributing to
the overall understanding and operationalization of the concept.

This accounts for both the qualitative definitions, the moral approach and the technical operationalization which are all an integral part of the conceptualization. But there still is some work to be done before we have the necessary guidelines for conducting a comparable analysis, even though a starting point is available.

The rest of the contribution will concentrate on the last item, i.e. the operationalization of the concept of sustainable development, and especially on the problem of how to measure the sustainability.

Elements of the operationalization of the principles

There are several crucial questions in connection with the idea operationalization of sustainable development and here the focus will be put on:

a)   How to get the right information

b)   How to do the registration of information, and

c)   How to transform this information into measurable and comparable and thereby into operational categories

The following considerations are primarily made with reference to the North Atlantic and the Circumpolar areas.

Getting the right information

An important question in connection with the problem of the operationalizing of the sustainable development is how to get the necessary information.

a) The traditional sources of information are *the national statistical agencies*. As there is usually no tradition for the collection of information concerning issues such as environment and resources, it will probably take quite a long time before the necessary preparedness has developed. In this connection there will be some marked differences in flexibility within the group of agencies:

One question concerns the size of the agencies as determinant for the flexibility. Usually: the larger the agency, the longer it will take.

Another question is how the registration works in the different countries. Denmark / the Nordic countries have a tradition for central registration of people, organizations and firms based on unambiguous identifications which makes it a rather simple case to expand the registration to cover environmental data of one kind or the other. But for many countries an unambiguous identification is considered a threat against the personal integrity, and here a registration of environmental information connected to individuals, organizations, firms, etc. implies the collection of information through surveys which is very time and resource-consuming.

Generally the information registered by the national agencies will be available for activities other than the ones they were originally designed for. In many cases the information complies with international standards, but often the national standards differ slightly from the international which causes some troubles when it comes to comparability.

b) The registration done by different *national ministries and other administrative boards* is varying from one country to another. Typical for many agencies such as ministries of environment, agriculture, fisheries, etc. is that their data collection is directed towards solving usually short-term administrative problems. Only to a limited extent they do a more general collection of information on a regular basis.

Due to the administrative character of the information it is usually not available for use outside the administrative system.

c) In connection with *research activities* done by *public research units* such as universities, other higher educational units, etc. the situation is rather comparable to the characteristics of the public authorities; data are collected in connection with the specific solving of problems or in search for specific answers, and usually they are also directed towards problems of short-term character. This means that they do not automatically secure a continuity or a comparability, neither in their definition nor in their registration.

Availability differs very much. Generally speaking the collected information is considered an integral part of the research projects and therefore not automatically available for other research activities[16].

Due to the widespread and still increasing research activities within international networks a great potential for common standards and comparability is in development.

d) In connection with *research activities* done by *private organizations* the problems are comparable to the ones mentioned above, except for the fact that the information collected is usually an integral part of the firm or organization, and therefore considered secret, at least at the time of collection.

e) Data collected by *international organizations* and *international research agencies* are usually meant to be used in international comparisons. Therefore they are often based on continuity and comparability, both in their definition and in their registration. On the other hand this is responsible for the limitations in supply and the time lag in availability, because it takes time to harmonize data from the different sources.

There are many potential sources of information, and if all the registered data were available and comparable the situation would probably allow for the immediate analysis of sustainability, no matter how it is defined. But due to differences in registration, classification and availability there will be a lot of work to be done before this is the case.

Due to the much greater flexibility of the smaller national statistical agencies they are probably the most efficient entry into the world of measuring sustainable development.

Registration of information

The traditional methods

The traditional methods of producing statistics at the national statistic agencies involved a lot of paperwork, manual registration and a division of labour between the producer of data, the developer of programs, and the producer of statistics.

These methods give severe limitations in connection with the production of alternative statistics because it usually takes a lot of work to transform the original data into the desired structures. And the ideas have to be passed through several levels of encoding/decoding before a result turns up, probably after a delay of several years.

Through the last years there have been several "cultural revolutions" within the statistic agencies due to the development in information technology, but even though it is possible to change the hardware and software and thereby create an up-to-date technical situation, the bureaucratic structures will remain for many years ahead keeping the traditional structure alive.

The great importance of data stemming from the national and international bureaus is the consistency of the information and the degree of documentation connected to it.

Information in relation to research activities

As the research activities are expected to be more up-to-date, they usually make use of more advanced equipment and tools, such as databases, advanced statistics, etc.

This does not only enable more sophisticated methods. What is much more important: It enables the interchange of large amounts of information between scientists, and allows the availability of vast amounts of data from more or less traditional sources of information, for instance in the form of CD-ROMs which are easily distributed, and can contain several hundred megabytes of data.

The main problem with data stemming from these sources is the limited consistency; they are of course are consistent within the relation they have been conceived, but they do not necessarily comply with the international standards. At the same time they are not necessarily well documented.

There is a great potential hidden in the huge amounts of scientific data - probably enough to enable the realization of the quantitative dimension of the sustainable development, if they could just be brought to a common comparable base.

Distributed information

And this is what the present development towards the distribution of information is all about[17]. Through the internationalization of the scientific networks the sharing of information does not only call for the possibility of interchange of information, but also for a high level of consistency and at the same time a high degree of documentation.

GIS

To be in a "state of the art" situation the information should be registered not only in a distributed information system but also in a GIS.

A GIS - Geographic Information System is a system which is able to handle not only the statistical dimension but also the spatial dimension of information.[18]

The importance of GIS in connection with the development of the concept of sustainable development could be formulated in this way:

"GIS technology represents the technological underpinnings for achieving environmentally responsible decision making. Achieving sustainable development requires the capability to pull information together and make it easily accessible for all interested parties.... GIS technology should help us attain an integrated view of the world, influence what we can and wish to achieve and, more significantly, determine the way we operate. [19]

The transformation of the information to operational categories

Accounting of matter

One way of keeping track of the development in environment and resources has already been introduced in part one. It is through a system of *accounting of matter* in which every level in the production (and reproduction) process is described by the amounts of elements, chemical compounds and energy passing through the system.

Advantages

The obvious advantages of this method are many. It is simple and easy to deal with because it uses the natural characteristics of the matter. It gives as mentioned before, the possibility of an exact tracing of all elements, and thereby the creation of a management tool which at the same time enables accounting of matter and energy, the optimization of processes, minimization of waste production, etc.

This system is, in the most simple form, already widespread in use in many countries, for instance in Denmark, as a basis for national taxation of CO_2 production and energy consumption.

More advanced systems such as SESAM (Sustainable Energy Systems Analysis)[20] which enable a complex evaluation of the consequences of for instance different policies for the energy consumption have been developed.

Disadvantages

Among the more marked disadvantages of this method is that it does not open up an evaluation of an alternative resource basis to a given production of an article for use. The method offers the possibility of evaluating different elements one by one, but not to compare them. As a planning instrument for the bureaucracy it is usable, but useless as a more dynamic instrument for the development of alternatives.

Accounting of human work

Another way of keeping track of the development is the *accounting of human work*. By this method every level of the production (and repro-duction) process is described by the amounts of human work necessary to produce and reproduce the elements of nature used in connection with the process.

Advantages

This way of accounting is also rather simple and easy to deal with because the objectives are well defined. At the same time the results are comparable.

Disadvantages

The most important disadvantage of this method is that a lot of important relations are not expressible in this category. How much human work does it for instance take to invent and develop an alternative basis for microelec-tronics?

Economic categories

As soon as one moves away from accounting matter and energy, the only real alternative turns out to be in economic categories.

As motion of matter and energy are expressible in human labour it is also convertible into other values.

Many authors, for instance Victor Anderson[21] or Herman E. Daly and John B. Cobb, jr.[22] try to elaborate on the traditional national income accounting as defined by John Maynard Keynes in 1936 .

According to Anderson the evaluation by means of the GDP lacks some important elements concerning the sustainable development. The following discussion is primarily based on Victor Anderson's book.

Step one: The "Green" GDP = ANP (Adjusted National Product)

One possible development is the modification of the GDP by creating an ANP - Adjusted National Product, for instance like this:

	ANP	=	GNP

(1)	- capital depreciation
(2)	+ money value of unpaid domestic labour
(3)	+ money value of non-money transactions outside the household
(4)	- environmental depreciation.

1) Capital depreciations should be subtracted. As it is now "a rapid 'throughput' of resources and goods will have a high GNP but without achieving a high level of ownership of goods at any time". This adjustment is already made in many situations and presented as NNP - Net National Product.[23]

45

2) *Money value of unpaid domestic labour* should be added because the work is the same whether it is done domestically or non-domestically.

3) *Money value of non-money transactions outside the household* should be added in order to reflect the real amount of work done.

4) Last, but not least, *environmental depreciation* should be subtracted in order to reflect the fact that resources have been reduced.

Step 2: The social indicators

The next step should include important social indicators in order to reflect how the ANP has been used in the development process. As indicators usable in global comparisons the following are suggested:

1) Net primary school enrolment ratio for girls
2) Net primary school enrolment ratio for boys
3) Female illiteracy rate
4) Male illiteracy rate
5) The rate of unemployment
6) Average calorie supply as a percentage of requirements
7) Percentage of the population with access to safe drinking water
8) Telephones per thousand people
9) Household income received by the top 20 per cent of households divided by that received by the bottom 20 per cent
10) Infant mortality rate
11) Under-five mortality rate.

Step 3: The environmental indicators

The third step should include important environmental indicators in order to reflect how the ANP has been used in connection with environmental problems. As indicators it is suggested:

12) Deforestation in square kilometers per year

13) Carbon dioxide emissions from fossil fuel use, in millions of metric tons per year

14) Average annual percentage rate of increase in population

15) Number of operable nuclear reactors

16) Energy consumption (in tons of oil equivalent) per million dollars of GDP (or ANP)

Advantages

One important advantage with the ANP is that it is accountable within the traditional national income accounting which makes its transition into the existing systems of accounting much easier.

The first step - the ANP - seems not only a reasonable extension of the GNP, but is also rather easily accessible.

The idea of supplementing the ANP with important indicators seems to be less operational, but on the other hand necessary to get a more balanced impression of the development process and its consequences.

Disadvantages

The supplemental indicators as specified might be interesting in a global comparison, but what they ought to be in for instance the evaluation of the development in the North Atlantic region or in the Circumpolar area is still to be discussed.

Other methods of measurement

Two other approaches to the measurement of sustainability should be mentioned.

One is the *Input-Output analysis* which in many ways resembles the systems of accounting of energy and matter by keeping track of the movements between different sectors.

The other is the *Cost-Benefit analysis* which raises the question of valuation of costs as well as benefits of a given development[24] in order to tell, not only quantitatively but also qualitatively, the consequences of this.

Examples of attempts to establish the statistical basis for implementation of sustainable development

To get an idea of how to get closer to the statistical basis for the implementation of the principle of sustainable development it is worth while to look at some examples of how it has been done so far.

First example: The OECD Environmental Performance Reviews[25]

OECD has decided to start a new programme of environmental monitoring in 1992. This implies that all members of the OECD have to conduct an analysis of the environmental conditions in order to improve it. The exact purpose is:

1) To reduce the overall pollution burden in the OECD countries,

2) To encourage the integration of the policies of environment and economics, and

3) To strengthen the co-operation with the rest of the world in matters concerning environmental questions.

The analysis should result in a report which, besides a general overview of the conditions of resources and environment, results in a number of conclusions and recommendations concerning the development of environmental initiatives. It is the hope from OECD that through their individual contributions the single members will prove to be innovative and farsighted.

All the members of the OECD will be analyzed in a certain cycle; every year 6 countries will be analyzed which means that every country will be analyzed at least every fourth year.

Iceland was the first country to be analyzed by OECD in 1992[26] according to these guidelines. The report comprises approximately 100 pages covering the following main themes:

1. The Context

Part I. Natural Resources Management and Pollution Reduction

2. Marine Resources
3. Terrestrial Resources
4. Pollution Control

Part II. Integration of Policies

5. Institutions and Instruments
6. Sectoral Integration and Nature Conservation

Part III. Co-operation with the International Community

7. International Commitments
8. Conclusions.

The report emphasizes (p. 10) that the first part is intended to be directed toward use in connection with sustainable development of the marine resources. As a part of the analysis the Icelandic quota system with individual transferability is emphasized as a good means to ensure sustainability. According to the report the control of the fisheries should be guided by *firstly* an attempt to achieve sustainable yield and conservation of fish stocks over the long term, using scientific criteria to limit total catches and to diversify the number of species caught. *Secondly* the objective should be a maximization of the benefits for the country by optimizing the economic efficiency of the fisheries, which primary means a reduction of the surplus investment and a cutting back on fishing effort. *Thirdly* it is mentioned that the means introduced should result in a differentiation in activities between the different regions, and at the same time encourage to full employment and a stable development of the population.

The concept of sustainable development is introduced, and it is undoubtedly a step in the right direction, but with a rather limited understanding of the concept of sustainability. It is difficult to see a massive difference between standard OECD reports and this one when it comes to recommendations.

Second example: The Icelandic report to UNCED

As an example of a new way of introducing the environmental consciousness the *ICELAND - National report to UNCED*[27] was presented at the conference in Rio in 1992. The report resembles in many ways an official status of the environmental conditions in Iceland which means that the presentation has been concentrated on items and subjects in which a number of national initiatives have been taken:

I. Development and Environmental Impact
1. Iceland
> The Settlement and Cultural Heritage of Iceland - Political Institutions
2. Environmental Endowment
> Climate - Geology and Soil - Physical Geography - Geothermal Activity - Land Use - The Marine Environment - The Ocean Resource
3. Economic and Social Development
> Population Growth and Demographic Trends - Economic Growth, Trends and Prospects - The Structure of the Economy - Distribution of the Labour Force - Government Welfare Services - Quality of Life - Energy Consumption

4. Development and Exploitation of Natural Resources
    Fisheries - Agriculture - Energy Resources and their Utilization - Industry - Tourism

II. State of the Environment and Natural Resources

5. State of the Atmosphere
    Background Air Quality - Anthropogenic Emission - Local Air Pollution - Noise Pollution - Toxic Residue in Living Organisms - Air-borne Radioactivity

6. The Ocean
    Status and Utilization of Marine Living Resources - Marine Pollution

7. Vegetation and Soil Resources
    Climate and Potential Vegetation - Deterioration of the Rangelands and Soil Erosion - Soil Pollution

8. Quality and Supply of Fresh Water
    Availability of Fresh Water for Human Consumption - Risk of Fresh Water Contamination

9. Biological Diversity
    Conservation - Flora - Wild Terrestrial Mammals - Bird Life - Freshwater Fish - Invertebrates - Reptiles and Amphibians

III. Conserving the Environment

10. Toward Sustainable Development
    Principles and Goals - The Ministry for the Environment - Environmental Legislation

11. Government Programmes and Projects
    Nature Conservation - Management of Ocean Resources - Soil Conservation and Afforestation - Physical Planning and Construction - Pollution Control - Environmental Education and Public Relations - Public Health and the Environment - Environmental Economics - Public Preparedness and Mitigation for Natural Disasters - Official Development Assistance

12. International Co-operation
    Global Collaboration - Regional Co-operation

13. Non-Governmental Organization

The analysis is primarily based on a verbal description, a few graphs and a number of standard tables. For a number of the most important parameters the environmental strain has been calculated:

Natural Resource Base
- The marine environment: basis for 70% of the export volume and 50% of export value; an average of 1.5 mil. tons - Agriculture : 3% of GDP and 5% employment - Hydro and geothermal resources : only 17% of the hydro resources are in use, and even less for the geothermal - Tourism and recreation : 4% of GDP, 9 % of foreign money, 5% of employment.

Energy consumption and supply
- 106 petajoules=417 gigajoules pr. cap. - Hydropower : 37% of total consumption -Hydroelectric power 93.5 % of electricity., geothermal 6.4% and diesel 0.1% - Geothermal energy : 31% of total, 85% of heating - Oil 29% and other fossils 3% of total; 85% of total transportation; Coal only in ferro-silicate, cement and alumina production

Vegetation and soil
- 80% of vegetation and soil disappeared since 1000 - Soil conservation - Afforestation - Pollution of terrestrial surroundings - Contamination of water from waste, sewage, (artificial) manure and pesticides.

Anthropogenous emission
- CFC gasses (0.02% of world consumption, 0.55 kg pr. cap.) - CO_2-emission (2.9 mil. tons = 790 GgC) = 11.3 tons CO_2 (3.1 tons C) per. cap. Of this 25% from transport, 25% from the fishing fleet, and industries 23%. - Methane = 20.000 tons - Nitrogenous oxides 1.200 tons - SO_2 11.000 tons (5.5 GgS) plus 8 GgS H_2S from geothermic energy. NO_x = 20.000 tons (as NO_2: 6 GgN)

Risks of marine pollution
- Eutrophication from sewage - Eutrophication due to aquaculture - Pollution from traffic

Governmental activities
- Participation in conventions: UN : Arctic Environmental Protection Strategy : The Nordic Environmental Programme : The Vienna Convention : The Montreal Protocol : The ECE Agreement : Oslo and Paris conventions

Third example: Statistics Canada

The third example is from *Human Activity and the Environment* published by Statistics Canada.[28] The main purpose of this contribution is to give a general statistical introduction to the man-nature interaction, not necessarily in order to establish a firm foundation for the evaluation of sustainability in Canada, but primarily in order to make the public aware of the overall situation as far as the statistics has been able to do by means of existing and available information.

The report consists of:

1. Preface
1.1   Introductions and Highlights

2. Population
2.1   Population Conditions
        Population and Population Density
2.2   Population Process
        Fertility, Mortality and Migration
2.3   Environmental Benefits and Impacts on Individuals
            Environmental Benefits - Impacts of Environment
2.4   Regulations and Perceptions
            Regulating Processes - Attitudes and Perceptions

3. Socio-Economy

3.1   Socio-economic Conditions

> Industry - Transportation and Utilities - Dwellings - Economic Indicators - Natural Resource Accounting

3.2   Socio-Economic Processes

> Production - Consumption - Mitigation

3.3   Environmental Impacts

> Contaminants - Agricultural Contaminants - Restructuring of Land and Water Systems - Composition and Range of Wildlife

3.4   Wastes and Recycling

4. Environment

4.1   Environmental Condition

> Air Quality - Water Quality - Fish and Wildlife - Habitats - Protected Land - Soil Quality - Agricultural Land - Forest Land - Minerals

4.2   Natural Processes

> Natural Disasters - Climate

4.3   Harvesting and Extraction

> Land - Fish and Wildlife Harvesting - Water Consumption

5. Appendices

5.1   The Population-Environment Process

5.2   Geographic Units for Environmental Analysis

5.3   Jock River Environmental Assessment: A case study.

The report could in many ways be considered as a general source-book of information concerning some of the problems related to environment much more than an attempt to pinpoint the crucial elements of sustainability. But especially the case studies presented can be considered an attempt to get a broader perspective on the problems of sustainable development, among others by pointing out the necessity of differentiating the analysis depending on whether it is on a local and on a regional or global scale.

Fourth example: The AMAP initiative

AMAP is the *Arctic Monitoring and Assessment Programme* with the primary objectives of measuring the levels of anthropogenic pollutants and assessment of their effects in the Arctic environment[29]. The programme should a) monitor, assess and report the status of Arctic Environment, b) document and asses the effects of anthropogenic pollution, c) recognize the importance and relationship to, and use of the Arctic flora and fauna by the indigenous people, and d) document the levels and trends of pollutants.

The programme consists of a number of separate monitoring activities:

- Monitoring pollutants
- Monitoring human health
- Monitoring effects
- Monitoring the atmosphere
- Terrestrial monitoring
- Freshwater monitoring and
- Marine monitoring.

The program should ensure that a co-operation among local and regional efforts and global programmes could result in a better documentation of the environmental situation in the Arctic. As such it is not directly a part of an initiative aimed at unfolding the principles of sustainable development on a national or international scale, but it is an example of an international activity necessary for ensuring the means of evaluation of the general environmental situation.

Strategy for the implementation of the concept of sustainable development in Greenland
The purpose of establishing the statistical basis for the evaluation of sustainability

There are several reasons to start the development of a proper basis for the evaluation of sustainability in Greenland, and for the evaluation of a "green" national product.

- Greenland will, within very few years, be confronted with the request from OECD to present a report about national environmental problems and initiatives related to the objectives described above. In order to be able to answer to this it is necessary to start as soon as possible.

- An initiative by the Nordic Council of Ministers concerning the development of a common basis for the national accounting considering environmental and natural resources i.e. a "green" GNP, to which an input from the members concerning specific national problems might be of great value.

- Statistics giving a thorough introduction to the situation of environment and resources in Greenland would be useful both in relation to education, research and in technical and administrative activities.

- In order to start the fulfilment of the ideas of sustainable development it is important that all countries which have engaged to observe the principles, start doing this as soon as possible; as the fulfilment requires both the technical basis and the broad acceptance of this way of thinking it has to be developed step by step, the first step being the establishing of the necessary information.

- As mentioned in connection with the problems of the sources of information Greenland is in a rather favourable position; The national agency of statistics - Greenland Bureau of Statistics - is small and flexible, a lot of national and international research activities are going on, and there is a possibility of establishing the necessary bureaucratic basis for harmonization and collection of data through *Den Videnskabelige Kommission for Grønland*. Not necessarily by presenting the scientists with a list of directives which should be fulfilled, but by putting the available info at the disposat of the scientists.

As mentioned in part 2 the activities concerning sustainable development consist of at least two steps:

- the establishing of the necessary statistical basis, and

- the synthesizing of a proper measurement of sustainability, i.e. the "Green" GNP.

The statistical basis is a prerequisite for the synthesizing of a measurement, and anyway for de evaluation of sustainability whether or not measured as a "green" GNP.

What to do

The first step is the establishing of a "Greenland Book of Resources" comparable to the one from Statistics Canada mentioned above, but with much more emphasis on the environmental issues and on the natural conditions and limitations of the fisheries and fishing industry.

The main purpose is to present, and generally make available, some of the information already existing inside the Greenland Home Rule Administration and in the different firms controlled by the Home Rule authority. The purpose is therefore that it should not be necessary to start collection of new information; it is primarily a question of making use of the existing.

The source of information should be the administrative units according to the following list:

Fiskeriundersøgelser	Fiskeressourcer (Fisheries)
	Havpattedyr (Sea mammals)
	Fiskeriregulering (Fisheries control)
Miljøundersøgelser	Landpattedyr (Mammals)
	Fugle (Birds)
	Ferskvandsfisk (Freshwater fish)
	Fredning (Preservation)
Forundersøgelser	Ferskvandsmålinger (Freshwater)
	Boringer (Drilling)
	Landskabsregistrering (Landscapes)
	Kortgrundlag (Mapping)
Energi	Energiforsyning og -forbrug (Energy - production and consumption)
	Vandforsyning og -forbrug (Freshwater - production and consumption)
Geologiske unders.	Geologi (Geology)
	Mineralske Ressourcer (Minerals)
Meteorologisk inst.	Klima (Climate)
	Isforhold (Sea-ice)
Farvandsdirektoratet	Strøm- og besejlingsforhold (Mapping of conditions in the sea)
Botaniske undersøgelser	Plantevækst (Plants)
Upernaviarsuk	Jordbund og erosion (Soil and Erosion)
	Planteproduktion (Plant production)
ØD-Fysisk planlægning	Kortplaner (Maps)
	Konfliktområder (Areas of conflicts)
GSK	Befolkningsdata (Population)
	Beskæftigelsesdata (Employment)
	Økonomi (Economy)
Sundhedsdirektorat	Sundheds- og sygdomsdata (Health)

First step: a Greenland Book of Resources

In order to cover as many as possible of the potential problems in connection with the discussions about sustainable development, the following preliminary outline of the Resource Book should be considered:

I   INTRODUCTION

II.   THE DEVELOPMENT PROCESS AND THE ENVIRONMENTAL CONSEQUENCES
A short introduction to how the development process in general changes the relations to nature, and an introduction to the concept of sustainable development

1. General introduction
Starting point:
Short introduction to the History of Greenland emphasizing the economic, social and cultural aspects of the transformation of nature
A short introduction to the history:
- The traditional production and relations to nature - Colonialism and relations to nature - Modernization and relations to nature - Greenland Home Rule and relations to nature
The political institutions of Greenland
A short introduction to the administration of Greenland

2. Environmental Fundamentals
A short introduction to some of the basic processes in nature

Climate
- Monthly Temperature and Precipitation - The changes in temperature and precipitation during the centuries - Ice-core analysis with a short introduction to the known history *(maps and tables from Met. Inst.)*

Geology and soils
Geology, Soils, Permafrost *(maps from GGU)*

Physical Geography

Quaternary maps - Landscapes *(maps from GGU, maps from Forundersøgerne)*

Land areas

Urban areas - agricultural areas - reindeer-areas - Muskox areas *(maps from Upernaviarssuk, miljøunders.)*

Flora

Distribution of plants *(maps/table from gbotanik)*

Terrestrial animals

Distribution of animals *(maps/table from miljøforv.)*

Birds

Distribution of birds *(maps/table from miljøforv.)*

Freshwater fish

Distribution of freshwater fish *(maps/table from gf)*

Invertebrates

Distribution of invertebrates *(maps/table from miljøunders.)*

The sea

Depth - Currents - Temperature - Ice *(maps from gf)*

Resources in the sea

Primary production - Species *(maps from gf)*

Energy

Energy production - Hydropower-potential - Wind power potential - Other energy resources / coal, oil, gas, *(table and graph from nunatek + gsk)*

3. The development of economic and social conditions

Population and demography

Demographic composition, growth, etc. *(table and graph from gsk)*

Economic development - tendencies

Economic key indicators; regional development *(table and graph from gsk)*

Economic structure

The key elements of the economy *(table and graph from gsk)*

Labour

Key elements of the labour force *(table and graph from gsk)*

Social security

Social security, Transfers *(table and graph from gsk)*

Quality of life

Hunting and fishing as spare-time activity *(table and graph from gsk)*

4. The development of resource management

Fisheries

Key data of fisheries - species - boats - areas - landings - regional development - investments *(table and graph from gsk)*

Agriculture

Production, Consumption, Labour, Waste, Economy *(table and graph from gsk)*

Energy - production and consumption

Production, consumption, Labour, Waste, Economy *(table and graph from gsk)*

Mineral and non-mineral resources

Production, Consumption, Labour, Waste, Economy *(maps from GGU)*

Industry

Production, Consumption, Labour, Waste, Economy *(table and graph from gsk)*

Tourism

Tourism - Activities, Labour, Environmental consequences, Economy *(table and graph from gsk)*

III.   *ENVIRONMENTAL CONSEQUENCES*

5. Consequences for the immediate surroundings
Day and night renovation

Activity, Environmental consequences *(maps from miljøunders.)*

Local consequences

        Smoke, Sewage *(maps from miljøunders.)*

Noise

        Noise *(Lufthavnsvæsnet)*

Airborne pollutants

        Radioactivity *(maps from miljøunders.)*

Working conditions

        Accidents, long term influences *(maps from sundheds-dir.)*

6. Consequences for the atmosphere

Air quality

        Gases, Dust, CFC *(maps from met?)*

Anthropogenous emission

        CO_2, NO_x *(maps from met?)*

7. Consequences for the sea

Status and exploitation of the marine resources

        (maps from gf)

Marine pollution

        (maps from gf/miljøunders.)

Poison in living organisms

        Concentration of poisoning elements *(maps from miljøunders.)*

8. Consequences for vegetation and soil

Climate and potential vegetation

        Potential production for sheep, reindeer, muskox *(maps from ror+gb)*

Erosion

        Erosion, decertification *(maps from upern.)*

Contamination of soils

        (?)

9. Consequences for supply of freshwater

Availability of freshwater

        Supply of freshwater *(maps+table from nunatek)*

Risk for contamination of freshwater

Oil *(maps + table from nunatek)*

10. *Consequences for the diversity*
Preservation
    (maps + table from miljøforv.)

IV.   ENVIRONMENTAL PROTECTION
    An introduction to the activities in this field
11. *Towards a sustainable development*
Principles and goals
    (table and text from miljøforv.)
Management of the environment
    Management principles, Institutions, Activities *(table and text from miljøforv.)*
Legislation
    (table and text from miljøforv.)
12. *The Home Rule programs for protection*
Protection principles
    Legislation *(table and text from miljøforv.)*
The sea
    Fisheries management, quoting, Legislation *(table and text from miljøforv.)*
The land
    Mammals, Birds - Management, quoting, legislation *(table and text from miljøforv.)*
Soil conservation and tree planting
    (table and text from miljøforv.)
Urban planning
    Principles, Conservation *(table and text from miljøforv.*
Pollution control)
    Measures *(table and text from miljøforv.)*
Public activities
    Education, Research, Other initiatives *(miljøforv.)*

Health

      (table and text from Parisaa.)

The economy of environment

      (table and text from miljøforv.)

How to be prepared

      A discussion about how to develop a preparedness

13. International co-operation

Global co-operation

      Participation in international organizations

Regional co-operation

      ICC activities, Nordic activities

14. Voluntary organizations

Voluntary organizations

      Dyrenes beskyttelse - Greenpeace - Danmarks
      Naturfredningsforening - other organizations

The second step: The implementation

It is impossible to make plans for the actual implementation of sustainability, but what can be considered is the implementation of the concept in connection with the evaluation of the economy. And as described in part 2 this could be done through the calculation of a "green" GNP.

Even though the intention of the resource-book is to cover most of the questions concerning man's relations to nature, the implementation of the concept of sustainable development will probably take a long time, a lot of trials, and possibly also a lot of errors.

In order to get started a few central elements could be tested, and a thorough selection would probably make the analysis accounting for the major part of the environmental costs.

The most important examples would probably be *Discard*. The problem of discard is undoubtedly the most important environmental problem in

Greenland today. The discard is attractive as long as the price structure and quota system is functioning as it is today. Some of the important questions are:

- The value of the discard - The consequences of discard for the stock development - The cost and benefit of alternative methods for selective fisheries - The cost and benefit of alternative methods of marketing.

Among other important issues which could be considered are:

Waste management: The cost of waste management and benefits of alternatives.

Energy consumption: The cost of energy consumption and reproduction - The cost and benefit of Alternative energy sources.

Building materials: The cost of reproduction of the sources of building materials - The costs and benefit of a national/local production of building materials.

By/Bygd: The cost and benefits of centralized versus decentralized settlement.

Notes

1. E.Goldsmith, R. Allen, M. Allaby, J. Davoll and S. Lawrence: A Blueprint for Survival. Special issue of The Ecologist. 1992.

2. A blueprint of Survival, Op. cit.2.21E. Goldsmith, R. Allen, M. Allaby, J. Davoll and S. Lawrence: A Blueprint for Survival. Special issue of The Ecologist. 1992.

3. Donella H. Meadows, Dennis L. Meadows, Jørgen Randers and William W. Behrens III: The Limits to Growth. Universe Books, New York 1972.3.21E.
    Goldsmith, R. Allen, M. Allaby, J. Davoll and S. Lawrence: A Blueprint for Survival. Special issue of The Ecologist. 1992.

4. Nogle fakta om den jord vi sammen bebor. NOAH 1971. 4.21E.
    Goldsmith, R. Allen, M. Allaby, J. Davoll and S. Lawrence: A Blueprint for Survival. Special issue of The Ecologist. 1992.

5. For instance in: Paul Elkins, Mayer Hillman og Rogert Hutchison: Verdens Velstand - En grundbog om grøn økonomi. Hovedland 1992. A book you could sympathise with because the intentions seem right, but generally without a clear reflection about the dynamic development of technology and thereby the relations to nature.
    Also the yearly publications from the Worldwatch Institute about The State of the World which has been produced since 1984 is representing this basic point of view.5.21E. Goldsmith, R. Allen, M. Allaby, J. Davoll and S. Lawrence: A Blueprint for Survival. Special issue of The Ecologist. 1992.

6. The Club of Rome has since Limits to Growth in 1972 published 17 other reports in which the discussions have been developed. The last one: The First Global Revolution from 1991 is focusing much more on themes concerning global inequalities, but has more or less abandoned the quantitative approach.
    Goldsmith, R. Allen, M. Allaby, J. Davoll and S. Lawrence: A Blueprint for Survival. Special issue of The Ecologist. 1992.

7. As an example could be mentioned: Sofus Christiansen: Subsistense on Bellona (Mungiki). Copenhagen 1982 or Ruthenberg: Shifting Cultivation, London 1976.

8. E. Neef: Zur Frage des gebietswirtschaftlichen Potentials. Forschung und Fortschritte; 40. Jg., 2/1966.
    E. Neef: Der Stoffwechsel zwischen Gesellschaft und Natur als geographisches Problem. Geographische Rundschau, 21, 12 1969.

H. Leser: Das Modell der Landschaft im Modell der Territorialstruktur; Landschaftsökologie. Stuttgart 1976.

J. Brandt: Landskab og Territorialstruktur. Nogle miljøgeografiske overvejelser omkring landskabsøkologisk teori.

9.   For a discussion of this concept of potential production, see :

R. O. Rasmussen: Naturressourcer, Økologisk Potentiale og Landbrug. København 1976.

R. O. Rasmussen: Potentialet og den samfundsmæssige produktion. Roskilde 1980.

R. O. Rasmussen: Informationsteknologi og informationsorganisering - konsekvenser for udviklingen indenfor fiskerisektoren. Roskilde 1992.

10.   J. Brandt and R.O. Rasmussen: Humanøkologisk Beskrivelsesmetode. RUC 1979,

11.   Aleksej Aleksandrovic Minc: Die ökonomische Bewertung der Naturressourcen. VED Hermann Haack, Leipzig 1976. Original Moskva 1972.

I.P. Gerassimow, L.S. Abramow, L.F. Kunizyn, N.F. Leontjew, J.G. Maschbiz, A. A. Minc, V.S. Preobrashenski: Mensch, Gesellschaft und Umwelt. Volk und Wissen Verlag, Berlin 1976. Original Moskva 1973.

12.   Autorenkollektiv: Gesetzmässigkeiten der intensiv erweiterten Reproduktion bei der weiteren Gestaltung der entwickelten sozialistischen Gesellschaft. Akademie-Verlag, Berlin 1976. In this especially chapter 8: Die Reproduktion der natürlichen Umweltbedingungen und die territoriale Wirtschaftsorganisation.

Hans Roos: Naturmæssige Miljøbetingelser og Nationaløkonomisk Reproduktionsproces. RUC Geografi 1978. Oprindelig trykt i Geographische Berichte, 80. årg. nr. 3.

Werner Gringmuth, Kurt Kutzchbauch, Hans Roos, Günter Streibel: Umweltgestaltung und Ökonomie der Naturressourcen. Verlag Die Wirtschaft, Berlin 1979.

Dokumente: Zur Materialökonomie. Verlag Tribüne Berlin 1978.

13.   Zur Materialökonomie: Dokumente, op.cit.

14.   Hans Roos 1976 and 1979 op.cit.

15.   World Commission on Environment and Development: Our Common Future, Oxford University Press, Oxford 1987.

16.   In Denmark there is an organization - DDA - Danish Data Archives - with the primary goal to collect larger collections of data resulting from research activities and make them available for other scientists.

17. A definition of Distributed Information can be found in: Rasmus Ole Rasmussen: Informationsteknologi og informationsorganisering - konsekvenser for udviklingen indenfor fiskerisektoren. RUC 1992.

18. The terminology System shows to the situation that at the same time it should be able to registrate, store, retrieve and manipulate information.

19. Ed B. Wiken, Paul C. Rump and Brian Rizzo: GIS Supports Sustainable Development. GIS World vol. 5 no. 10, 1992, pp 55-57. The analysis is based on:

M.E. Colby: Environmental Management in Development: the Evolution of Paradigms. World Bank Discussion Papers No. 80, Washington, D.C., 1990.

J.D. Collision and E. Wiken: The State of Canada's Environment. Ottawa 1991.

J. T. Mathews: Moving Toward Eco-development: Developing Environmental Information for Decision Makers. Government of Canada, 1992.

Proceedings: Environmental Information Forum. International Forum on Environmental Information for the Twenty-First Century, Montreal, May 21-24, 1991. p-22-26.

20. A large number of reports from the SESAM project have been published. Some crucial elements have been presented in:

Klaus Illum: Bæredygtige Energisystemer - Problemstillinger og Modeldannelser. Sesam-skriftserie nr. 1, Ålborg Universitetsforlag,- Januar 1991.

Klaus Illum: Bæredygtige Energisystemer - En studie i Storstrøms Amt. Sesam-skriftserie nr. 2, Ålborg Universitetsforlag, Februar 1991.

Klaus Illum og Henning Mæng: Dokumentation af beregningsforudsætninger og resultater. Ålborg Universitetsforlag, August 1992.

21. Victor Anderson: Alternative Economic Indicators. Routledge, London 1991.

22. Herman E. Daly & John B. Cobb, Jr.: Det fælles bedste - En økologisk økonomi for fællesskab og fremtid. Hovedland 1991 (For the Common Good, Beacon Press 1989)

23. Daly and Cobb instead suggest the introduction of the Hicks-income, named after Sir John Hicks who stressed the importance of having a definition of national income which reflects how much it is possible to spend without making onself poorer. The Hicks-Income HI is defined

as:

HI = NNP - DE - DNR

where

NNP = Net National Product

DE = Defensive expenses

DNR = Depreciation of the natural resources.

24. For a discussion of this, see forinstance: David Pearce and Kerry Turner: Benefits estimates and Environmental Decision-making. OECD, Paris 1992.

25. OECD Environment Directoreate: Environmental Performance Reviews - Detailed Plan and Implementation Strategy. Paris 1992.

26. OECD Environment Directorate: Group on Environmental Performance: Environmental Performance Review of Island. Paris 1992.

27. Ministry for the Environment: ICELAND - National report to UNCED, Ministry for the Environment 1992, Reykjavik

28. Statistics Canada: Human Activity and the Environment; Minister of Industry, Science and Technology, Statistics Canada, Ottawa, 1991

29. Arctic Monitoring and Assessment Programme (AMAP). Draft proposal for the content of AMAP. Oslo 1992.

References

Anderson, V., 1991, *Alternative Economic Indicators*, Routledge, London.

Arctic Center Publications no.2, 1991, The State of the Arctic Environment. Arctic Center University of Lapland, Rovaniemi.

Arctic Monitoring and Assessment Programme (AMAP), 1992, Draft proposal for the content of AMAP. Oslo.

Autorenkollektiv, 1986, *Gesetzmässigkeiten der intensiv erweiterten Reproduktion bei der weiteren Gestaltung der entwickelten sozialistischen Gesellschaft*. Akademie-Verlag, Berlin.

Bartelmus, P., 1986, *Environment and Development*. Allen and Unwin, London.

Birou, A., Henry, P-M., Schlegel, J. (Eds), 1977, *Toward a Re-definition of Development*. Pergamon Press, Oxford.

Brandt, J., 1991, *Landskab og Territorialstruktur - Nogle miljøgeografiske over-vejelser omkring landskabsøkologisk teori*. RUC.

Brandt, J., Rasmussen, R.O., 1979, *Humanøkologisk Beskrivelsesmetode*. Forskningsrapport nr. 4, Institut for Geografi, Samfundsanalyse og Datalogi.

Brown, Lester R. et al, 1991, *Verdens Tilstand 1991 - State of the World 1991*. Worldwatch Institute Rapport. Mellemfolkelig Samvirke, Skive.

Colby, N.E., 1990, *Environmental Management in Development: the Evolution of Paradigms*. World Bank Discussion Papers No. 80, Washington, D.C.

Collision, J. D. and E. Wiken, 1991, *The State of Canada's Environment*. Ottawa.

Commoner, B., 1967, *Science and Survival*. Viking Compass Edition, New York.

Coomer, J.C. (Ed), 1979, *Quest for a Sustainable Society*. Pergamon Press, Oxford.

Christiansen, S., 1982, *Subsistence on Bellona* (Mungiki). Copenhagen.

Daly, H.E. and Cobb, J.B.jr., 1989, *For the common good*. Beacon Press, Boston.

De la Court, Thijs, 1990, *Beyond Brundtland*. Green Development in the 1990s. Zed Books, London.

Dokumente, 1978, *Zur Materialökonomie*. Verlag Tribüne Berlin.

Elkins, P. , Hillman M. and Hutchison R., 1992, *Verdens Velstand - En grundbog om grøn økonomi*. Hovedland.

Enyedi, G., Gijswijt, A., Rhode, B. (Eds), 1987, *Environmental Policies in East and West*. Taylor Graham, London.

Gabor, D., Colombo, U., King, A. , Galli, R., 1981, *Beyond the Age of Waste*. Pergamon Press, Oxford.

Gerassimow, I. P., L.S. Abramow, L.F. Kunizyn, N.F. Leontjew, J.G. Maschbiz, A. A. Minc, V.S. Preobrashenski, 1973, *Mensch, Gesellschaft und Umwelt*. Volk und Wissen Verlag, Berlin 1976. Original Moskva 1973.

Goodland, R., El Serafy, S. (Eds), 1991, *Environmentally Sustainable Economic Development: Building on Brundtland*. Unesco, Paris.

Gringmuth, W., Kurt Kutzchbauch, Hans Roos, Günter Streibel, 1979, *Umweltgestaltung und Ökonomie der Naturressourcen*. Verlag Die Wirtschaft, Berlin.

Hamrin, R.D., 1980, *Managing Growth in the 1980.s*. Praeger, New York.

Illum, K., 1991, *Bæredygtige Energisystemer - Problemstillinger og Modeldannelser*. Sesam-skriftserie nr. 1, Ålborg Universitetesforlag, Januar 1991.

Illum, K., 1991, *Bæredygtige Energisystemer - En studie i Storstrøms Amt*. Sesam-skriftserie nr. 2, Ålborg Universitetesforlag, Februar 1991.

71

Illum, K. og Mæng, H., 1992, *Dokumentation af beregningsforudsætninger og resultater*. Ålborg Universitetsforlag, August 1992.

King, A., Schneider, B., 1991, *The First Global Revolution*. Pantheon Books, New York.

Leser, H., 1976, *Das Modell der Landschaft im Modell der Territorialstruktur*; Landschaftsökologie. Stuttgart.

Marston, A. (Ed), 1992, *The Other Economy - Economics Nature can live with*. Learn By Doing Publishers, Auckland.

Mathews, M.T., 1992, *Moving Toward Eco-development: Developing Environmental Information for Decision Makers*. Government of Canada.

Meadows, D., Meadows, D., Randers, J., Behrens III, W. W., 1972, *The Limits to Growth*. Universe Books, New York.

Minc, A. A., 1972, *Die ökonomische Bewertung der Naturressourcen*. VEB Hermann Haack, Leipzig 1976. Original Moskva 1972.

Ministry for the Environment, 1992, *ICELAND - National report to UNCED*, Ministry for the Environment, Reykjavik.

Neef, E., 1966, Zur Frage des gebietswirtschaftlichen Potentials. *Forschnungen und Fortschritte*; 40. Jg. 3/1966.

Neef, E., 1969, Der Stoffwechsel zwischen Gesellschaft und Natur als geographisches Problem. *Geographische Rundschau*; 21, 12/1969.

Novikov, Yu., 1990, *Environmental Protection*. MIR Publishers, Moscow.

OECD, 1992, *Environmental Performance Reviews - Detailed Plan and Implementation Strategy*. Paris.

Pearce, D. and Turner, K., 1992, *Benefits estimates and Environmental Decision-making*. OECD, Paris.

Proceedings, 1991, Environmental Information Forum. *International Forum on Environmental Information for the Twenty-First Century*, Montreal, May 21-24, 1991. p-22-26.

Rasmussen, R.O., 1976, *Naturressourcer, Økologisk Potentiale og Landbrug*. Geografisk Institut, Københavns Universitet.

Rasmussen, R.O., 1980, *Potentialet og den Samfundsmæssige Produktion*. RUC.

Rasmussen, R.O., 1992, *Informationsteknologi og informationsorganisering - konsekvenser for udviklingen indenfor fiskerisektoren*. Roskilde.

Roos, H., 1980, Naturmæssige Miljøbetingelser og Nationaløkonomisk Reproduktionsproces.*Geographische Berichte*, 80. årg. nr. 3.

Ruthenberg, 1976, *Shifting Cultivation*, London.

Smith, E. (Ed), 1991, *Sustainable Development Through Northern Conservation Strategies*. The University of Calgary Press, Toronto.

Statistics Canada, 1991, *Human Activity and the Environment*; Minister of Industry, Science and Technology, Statistics Canada, Ottawa.

Stockholm Group for Studies on Natural Resources Management, 1988, Perspectives of Sustainable Development. *Stockholm Studies in Natural Resources Management* No. 1, Stockholm.

The Ecologist, 1972, A Blueprint for Survival. *The Ecologist* Vol.2, No.1, London.

Wiken,B. Paul C. Rump and Brian Rizzo (Eds), 1992, GIS Supports Sustainable Development. *GIS World* vol. 5 no. 10, 1992, pp 55-57

World Commission on Environment and Development, 1987, *Our Common Future*, Oxford University Press, Oxford.

Sustainable Development in the Arctic: Inherent Problems and Possible Solutions in a Theoretical Perspective

Urban Ignaz Hügin

The aim of this article is to point out some general aspects of inherent problems in »sustainable development« from a theoretical perspective (chapter 1) and to elaborate some possible guidelines for solutions (chapter 2).

1. Inherent Problems

»Sustainable development« implies the aspect of planning the future. However, planning the future of human systems (such as social, economic, political or ethnic groups) is extremly difficult. This is due to the fact that human systems are *open* (1), *non-linear, recursive* (feedback) systems (2) and with great sensitivity to *initial conditions* (3).

1. A human system is *open* because it is connected with its ecological, economic, social, political and cultural environment. Contextual factors influence changes and innovations in human systems to a high degree.
2. A human system is *non-linear* and *recursive* (feedback) because the actual reality of the human system at a given moment (m_1) determines the reality at a later moment (m_2) which in its turn determines the reality at an even later stage (m_3), etc.
3. The great *sensitivity to initial conditions* becomes evident by the fact that minute initial differences in two more or less identical systems can lead to complete different behaviour and situations in the long run, even when outside influences are the same.

These three interdependent facts can be illustrated by means of the mathematical equation below (cf. Fig. 1). The equation is known as the »logistic equation« (cf. GEO WISSEN: 184 - 185) and has been applied in many different fields. Here it serves as an analogous model for illustration.

Figure 1: Mathematical Model

$$x_{t+1} = rx_t (1 - x_t)$$

Figure 2: Numerical Application

| | r=2 | | r=4 | |
| | $x_0=0,4$ | $x_0=0,2$ | $x_0=0,4$ | $x_0=0,401$ |

Column :	I	II	III	IV
x_1	0.48	0.32	0.96	0.960796
x_2	0.4992	0.4352	0.1536	0.1506681
x_3	0.4999	0.4916	0.5200281	0.5118688
x_4	0.5	0.4998	0.9983955	0.9994365
x_5	0.5	0.5	0.0064077	0.0022527
x_6	0.5	0.5	0.0254665	0.0089905
x_7	0.5	0.5	0.0992718	0.0356386

We assume that a human system obeys deterministic laws and is describable by means of the »logistic equation«. In our case, the equation describes the dynamics and the development of a human system x over time t. Let x be a complete description of a given human system at a given moment t, for example a village in the North in 1992. The right-hand side of the equation $(rx_t \{1 - x_t\})$ describes all the connections, relationships and interdependences in and of this human system. r is the factor which describes the influence on the system from the outside.

As can be deduced from the numerical calculations in Fig. 1, three *interdependent* things are evident:

1. Factor r determines to a very high degree the long-term run of our system. Even minute differences in r can lead to completely different dynamics of the system (cf. Column I and Column III).
2. On one hand, initial differences between systems can diminish with time as in Columns I and II. On the other hand, minute differences in initial internal conditions of two almost identical systems can lead to growing differences with time (cf. Columns III and IV). After only seven steps, the initial difference has already grown over 250 times in scale).
3. Because every value of x at a time moment t depends on the preceding value of x, the individual »history« of the equation plays a crucial part in the dynamics.

In light of these facts, we can, at this point, make the following conclusions with respect to the problem of »sustainable developemt«:

1. The *external influence* on a system determines its dynamics to a very high degree.
Additionally, these external influences vary in quality and quantity with time and are never completely controlable by the system itself.
2. The *non-linearity* and *recursive feedback* of the human system cause uncertainty in predictions. Therefore, one cannot plan the state of a future moment in time, x_7, by a simple extrapolation of the present moment, x_1 (as in a linear system), but must take into consideration the reality of the future moment, x_6, and of course, the ever changing influence from the outside. Even when deterministic laws are available to help plan the future — an assumption which can be postulated epistemologically and methodologically, but which can never be accomplished empirically — the concrete dynamics of a human system in the future can never be predicted for a long time frame.

3. Due to the *sensitivity* of a human system *to initial conditions*, we must infer that what proves to be a good and successful plan in one case, is not necessarily such for an almost identical case.

In conclusion: One can never be sure that the development of the present does not compromise the ability of future generations to meet their own needs, even when one has the ethical intention not to do so. This aspect cannot be considered seriously enough. In light of these facts, planning of sustainable development in the Arctic seems impossible.

2. Possible Solutions

The discussed aspects are but one side of the coin. In other words, the situation for planning sustainable development is not as bleak as it may seem at a first glance. Why? Because human beings and human systems are not only connected with the past, as is expressed in the »logistic equation«, but are bound to the future as well. They act in aspiration and react in anticipation of the future.

Understanding the basic connections, relationships and mechanisms of human systems may contribute to finding ways for sustainable development in the Arctic as well as for other regions of the world. It is the aim of this chapter to point out *one* possible theoretical framework which might help to solve the problem of sustainability in a rather general manner. This will be attempted by developing the theoretical model as shown in Figure 2.

2.1 Theoretical Model

The theoretical model is mainly based on the structural theory of synergetics.

Synergetics is defined as the science of collective phenomena in multi - component, open systems with »cooperative« interactions occuring between the elements of the system (c.f. Haken 1980).

Synergetics is therefore applicable to the description and the understanding of the dynamics of the individual humans within a group as well as the dynamics of the whole group.

According to the synergetic theory a human group is *open* because there exists not only an internal interaction of material and immaterial nature between the members of the group, but also interaction with the external environment (e.g. other human groups or the physical environment). A human group is a *multi - component system* because it is composed of »units«, namely its individual members. Although these members can adopt different individual attitudes and behaviour, they behave in a more or less regular, similar manner. This is the result of common conventions about behaviour and attitudes in societal contexts. These common conventions are the results of the *cooperative interactions* between the individuals of the society and are a necessity for a collective existence.

There are three main *components* in this theory which describe a human system:

Component 1:   The »attractor«.
Component 2:   The »order-parameters«.
Component 3:   The »control-parameter«.

Every system has an »*attractor*«. The attractor describes the stable or ideal state of a system. For human groups it is »sense« or »meaning« that forms the attractor. The members of a human group try - individually and collectively - to organize the world they live in in a logical, coherent and consistent manner. Such a world makes sense and is meaningful. If inconsistencies exist or arise, »sense« or »meaning« becomes deficient and the system becomes instable.

The dynamics of systems are determined by the so-called *»order-parameters«*. Order-parameters contain the information necessary for the description of the dynamics of a system. In this context, the dynamics of human groups are determined by four interdependent »order-parameters«, namely:

— The *elements of the concrete reality* (e.g. an economic system, a social organization, or a simple fact like a bicycle).
— The *elements of the abstract reality* (e.g. values, norms, or more generally, the world view).
— The process of *differentiation*.
— The process of *integration*.

The two notions »differentiation« and »integration« have to be clarified briefly: A *differentiation* describes the process of change of the existing concrete and abstract reality in a human group. This can be an innovation in the form of a change of a single element, or the introduction of a new element from the outside world, etc. Thus, technical, economic and social aspects can be differentiated as well as aspects of the world view. *Integration* describes the process of how the elements of change are added and incorporated into the existing world. This is done by means of elements of the respective world view such as values and norms. More generally, integration is done by hypotheses which connect different elements to a coherent totality.

The last important variable is the *control-parameter*. It can be understood as the quantity and quality of the input into the system from the outside, the factor r in the »logistic equation« in Fig. 1.

With these three components — the attractor, the order-parameters and the control-parameter — it is possible to construct a qualitative model of the dynamics of human groups as presented in Fig. 2.

Figure 3:

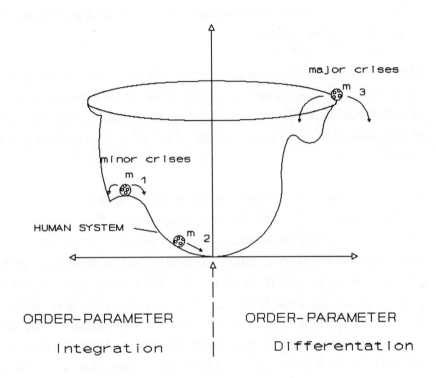

CONTROL–PARAMETER

Input

major crises

m 3

minor crises

m 1

HUMAN SYSTEM

m 2

ORDER–PARAMETER

Integration

ORDER–PARAMETER

Differentation

ATTRACTOR

"Sense","Meaning"

A human group can be imagined as a ball in a bowl, where the bowl represents the concrete and the abstract reality for the group. The input from outside works as an impetus which affects the ball. Depending on the degree and the quality of the impetus, the ball rolls from one point in the bowl to another. Some inputs result in the direction of differentiation, others in the direction of integration. Corresponding processes of differentiation and the respective processes of integration result in a partial transformation of the bowl. Thus, the bowl continuously changes its form as a consequence of an ever changing reality.

The ball always tends towards the bottom, seeking the most stable state for itself in the bowl (m_2). Because a time lag always exists between the process of differentiation and the process of integration, minor crises (m_1) are a normal phenomenon in human systems. The ball is but rarely in its ideal state.

With a sufficiently strong input on the system from the outside, the ball reaches the edge of the bowl. In the position of m_3 in Fig. 2, the input has led to a differentiation to such a degree that the possibility of the integration into the existing context has become questionable. The ball is in an entirely instable state. If the next input leads to integration, the ball will return back into the bowl. If the input leads to yet more differentiation, the ball will leave the bowl. This would mean the »breakdown« of the entire system and is considered a *major crisis*. Outside the bowl, the ball would roll back and forth, right and left in a chaotic manner, seeking by trial and error a new, bowllike vessel with an attractor that gives »sense« and »meaning« in the new situation.

As already mentioned, human goups are anxious to live in a meaningful world. Through self-organization human groups try to influence and direct the processes of differentiation and integration in such a way that a change makes sense. Therefore they try to select the input into their system in anticipation of possible problems and act in the spirit of fulfilment of the aspired future.

2.2 Possible Guidelines for Solutions

How can such a model be applied to the problem of sustainable development? Some general conclusions can be deduced which might help the persons who tries to plan and implement a sustainable development in the Arctic.

1. The concept of »sustainable development« has been defined as »a development which meets the needs of the present without compromising the ability of future generations to meet their own needs«.(Brundtland report). This concept focuses not only on ecological crises but stresses the importance of avoiding economic crises as well. Because only a holistic view of human nature and behaviour makes it possible to meet the challenge of »sustainable development«, avoidance of social and mental crises should be included in this concept.

2. »Sustainable development« means, with use of the model that the ball never leaves the bowl even when large changes occur. But one must take into consideration that crises are a normal phenomenon of human systems. Therefore avoidance of crises can not be a question of principle, instead it is one of degree.

3. A crisis can easily be caused by external influences which cannot be controlled by the influenced human group. It is therefore imperative that the control and the decisions about changes and innovations for sustainability can be made as far as possible by the affected group.

4. It is important to note that a process of integration must not be preceded by a process of differentiation. A value or norm can be an integrated part of the existing abstract reality of a human group and force a differentiation of the concrete reality so that »sense« and »meaning« is created in the system. It is therefore of great importance that »sustainability« be considered as an ethical value which creates »sense« and »meaning« in every human of the affected group. If this is the case the group will act and react in such a manner that the integrated value of »sustainability« is fulfilled continuously.

83

5. Since reality is an ever changing entity, the concrete form of »sustainability« must be understood as a contextual changing adaptation to an ever changing world and in anticipation of the possible future. Thus, while the abstract notion of »sustainability« must persist as an ethical value, the concrete forms that »sustainability« take on, have to be continuously adapted. This means open planning for the future.

3. Concluding Remarks

If the human race is seen as worthy of existence, sustainability is an absolute necessity. On the other hand implementing sustainable development is an extremely difficult task in our turbulent world today. But even when the efforts are not successful, the will to change something is a great merit, or, to say it in the ancient roman way:

Ut Desint Vires Tamen Est Laudandum Voluntas.

4. References

Wissen, Geo, 1990, *Chaos + Kreativität*, Hamburg

Festinger, L., 1978, *Theorie der kognitiven Dissonanz*, Bern, Stuttgart, Wien.

Haken, H., 1980, Synergetics. Are cooperative Phenomena governed by universal Principles?, In *Naturwissenschaften*, 1980 (67): 121 - 128.

Willke, H., 1982, *Systemtheorie*, Stuttgart, New York.

Cultural Sustainability - Anthropological Perspectives on Terrestrial Animal Production Systems in Greenland

Hans-Erik Rasmussen

Introduction

The present economic and ecological problems in Greenlandic fishery (especially cod and shrimp) sharpens the need for new directions in the utilization of renewable resources. In this connection it is necessary to discuss more closely a range of conditions surrounding the general cultural mechanisms which are associated with the exploitation of animal resources. The current and serious problems require that we pay close attention to the following problem areas:

A. Marine resources

(1) the possibility of identifying new types of fish for whole or partial replacement of cod and shrimp;
(2) the possibility of developing and/or improving the technology associated with the fishery sector in such a way that greater flexibility is achieved both with respect to the type of catch and with respect to the possibilities of landing the catches at various land installations;
(3) the possibility of developing various types of aqua-cultures.

B. Terrestrial resources

(1) the possibility of intensifying and/or expanding production systems already functioning (that is sheep breeding and reindeer herding);
(2) the possibility of introducing new production systems (for example muskox farming and yak farming).

Animal product processing has an ecological, an economic, and a cultural dimension. And these dimensions have both a local or national and a global perspective. With respect to Greenland the most important ecological factor is the Arctic ecosystem's vulnerability and high degree of specialization. And the most important economic factor is the fact that the all-prevailing part of Greenland's own economy and own business community build on various forms of exploitation of wildlife animals.[1]

Which animals are exploited, and the way in which they are exploited is not just a question of qualifications specific to a profession, but to a high degree a part of the cultural competence. The animal-based economy constructs and maintain cultural and ethnic identity.[2]

In this paper I shall concentrate on the terrestrial resources - not because the productivity in this case is largest, [3] but because its in the context of this type of resource exploitation that the *cultural* problems are most prominent. It becomes steadily clearer that models of resource exploitation and management which only consider the ecological and economic parametres are not sufficient for ensuring sustainable exploitation. This observation is valid whether it is desired to try to establish a greater national degree of self-sufficiency or whether it is desired to continue the present extreme export-dependency.

Selected issues

The introduction of production systems using "domesticated" animal species is known in almost all of the Inuit-populated area. Reindeer herding was brought to Alaska for the first time in 1891-92, to Canada in 1935, and to Greenland in 1952. Muskox farming was established for the first time in Alaska in the 1930's, and more systematized sheep breeding began in Greenland in 1915. Despite the relativily great regional differences, differences in economic and political history, etc., a range of problems with common traits has arisen over the preceding 100 year period. By systematizing and comparing the experiences from Alaska, Canada and Greenland it is easier to avoid a number of the conflict situations associated with possible establishment of new animal production systems in Greenland.

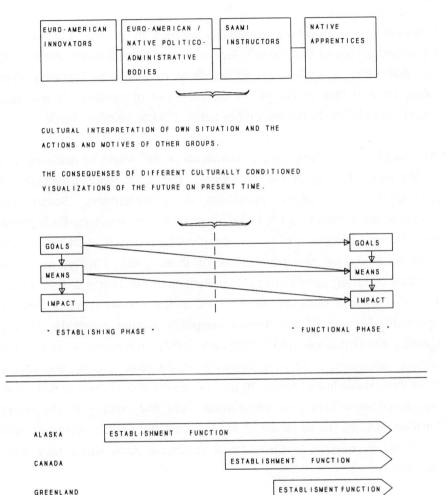

Figure 1

The scope of the present paper can only allow for a simplified presentation of the whole problem complex, but I shall attempt anyway to put forward a few central aspects. A number of the most important variables involved in the production systems associated with the discussions in the paper follow: Figure 1.

Common to the inception of reindeer herding in Alaska, Canada, and Greenland is the fact that the initiative was taken by non-native innovators, and that Saami specialists (especially from Finnmark in Norway) were called upon to take charge of the establishment of herding.[4] It was also Saami who undertook the training of native Eskimo reindeer herders.

An "establishment phase" and a "functional phase" might be made explicit in the case of each of the operations. The four major categories of participants - innovators, politicians & administrators, Saami and Eskimos/Greenlanders - each had their goals for the projects. Each group has different conceptions of which operational and organizational resources were necessary for the achievement of these goals. Finally, each had different culturally conditioned interpretations of function and disfunction. It is also difficult to obtain an overview of the various reindeer herding operations because of the diverse temporal starting points in Alaska, Canada, and Greenland (1891, 1935, and 1952). In the case of each of the projects internally, there were influences in the goals, means, and impact from the "establishment phase" to the "functional phase", just as there also was a certain influence on these areas from one country to the other. Furthermore, the native modes of production differed with respect to each other and in relation to the colonial structures upon which they were dependent.

In my opinion, a number of the most important problems associated with the animal production systems introduced can be summarized in the following way:

1. Social stratification.
The influence on and development of intra-ethnic social inequality in the local community;

88

2. Ethnic stratification.

The cultural consequences of the fact that reprensentatives from different ethnic groups took part in superordinate and subordinate relations as innovators, "teachers", apprentices, "recipients", etc.;

3. Regional differences.

The influence on and significance of already existing intra-ethnic regional cultural differences;

4. Economic and ideological structures.

The influence on identity-making and identity-maintaining relations between animals and human beings;
The influence on proprietal and distributional relations.

As will be seen, all of these points touch on *socio-cultural* aspects of animal resource exploitation.

Goals in animal production

The conception of whether the individual animal production projects function or do not function is naturally dependent on the goals set for the project. The external reason for the introduction of animal production systems in Alaska, Canada, and Greenland was, we are told, a falling sustenance base due to a reduction in the wild reindeer herd (Stern 1980:85, Naylor et.al. 1980:251-52; Seguin 1991:8; Rasmussen 1907; De Forenede Grønlandske Landsråds Forhandlinger 1946:631-33, 1947:763-65, 1948:40+60-64).[5] The purpose was therefore to create a stable food supply. The introduction of animal production systems in the three countries can be seen as a part of a broader professional development programme which presupposed that a large part of the native population continued to support themselves on hunting and sealing. Reindeer herding came to Greenland after the start of a strategy for the buisiness life which build upon the introduction of sheep breeding and the new development of fishing in coastal areas.

89

But the introduction of animal production systems had another equally important, but hidden agenda: civilizing the "primitive" hunter population (Stern 1978:150; Stern 1980:98; Seguin 1991:8-9). In the minds of the nonnative white innovators, hunter cultures with their "vagabond" populations represented the most primitive stages in cultural development. According to this opinion *pastoral* cultures lay on a higher "level of civilization" due to the close bond between the pastoralists and their "domesticated" animals. A negative aspect in this connection was, however, that nomads were not sedentary! The Euro-American conception of the ideal civilization was represented by the sedentary (and maybe piestistic) farmer.

It was therefore in keeping with this cultural view that Saami pastoralists were called in as instructors in connection with the plans for the establishment of Eskimo reindeer herding in Alaska, Canada, and Greenland. Nomadic *intensive* reindeer herding was the essence of the Saami conception of the ideal civilization. This idea, therefore, lay behind the Saami evaluation of to which extent Eskimos/Greenlanders mastered reindeer herding. (Se also the excellent discussion of the situation in Alaska in Beach 1986:53 ff).

The aspect of civilizing also appears explicitly in the Danish discussions which preceded the more systematic introduction of sheep breeding in Greenland. There were, nevertheless, people who doubted that Greenlanders were sufficiently civilized and enlightened to start up experimental farming and animal husbandry! (Anonymous 1906:42-66).

The agenda in animal production changes with time, and such goals of civilizing are no longer actual - not even as a hidden agenda. But what is problematic in this context is that the ethno-centric opinions behind the former goals can also be recognized in certain administrators of our time, in their unofficial interpretations of the cause of the actual problems in Greenlandic reindeer herding. But let us now turn to the present situation and a few, but very central issues.

A modern goal for animal production in Greenland could be *to increase the national degree of self-sufficiency in food supply*. At the same time, production should be ecological, economic *and* culturally sustainable - in other words, it should be able to propagate cultural traditions and form a basis for the maintenance of a culturally and ethnically consistent livelihood. We can also say that production should meet the criteria for *biological profitability:* net demographic growth; *economic profitability:* net economic growth; *cultural profitability:* accumulation of social status.

The goals sketched above should be met within the confines of a more superordinate goal of preserving the biological, commercial, and cultural diversity. Commercial diversity can mean a plurality of *specialized* economies, but the diversity can also include different forms of *mixed economies*. There is a tradition for exactly mixed economies as such in the Greenlandic household economy. In that it meets the mentioned criteria, animal product processing should, thus, be able to make up a part of such mixed economies.[6]

A multifacetted production strategy can be sketched in the following way:

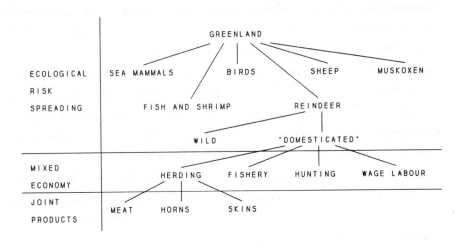

Figure 2

91

A basic decision *at the national level* must be made on the relationship between production for export and production for the domestic market. At *the level of the producer* a decision must be made on the relationship between trade and subsistence production. As a form of mixed economy, one can also imagine animal production involving *compound* stocks, e.g. reindeer, sheep, and muskoxen. As part of "risk spreading", one could utilize the animal's different grazing behaviour and growth rates.

Means and consequences in animal production

The goal of high self-sufficiency has various consequences depending on which level of society is under discussion. Politico-economic contrasts can easily arise between the different levels: A high degree af self-sufficiency on the *national* level presupposes a high degree of domestic trade (Danish: indhandling) which leads to a "redistribution" via the market (despite the specific price policy). On the other hand, a high degree of domestic trade undercuts the possibility of a high degree of self-sufficiency at the *level of the household*.

Economically, domestic household production can only be maintained in the context of cash incomes from wage-labour, the sale of products, etc. (Dahl 1989). The "cultural compensation" for wage-labour generally consists of the possibility of sealing, hunting and fishing at other times in the household cycle. With respect to "the hunter household" it would, therefore, be artificial to separate domestic and commerical production, precisely because subsistence economy and cash economy are in this case two sides of the same issue (Ibid.). This context, known from the Greenlandic hunting districts can only with difficulty be transferred to, for example, *reindeer herding*, unless this is very extensive. Reindeer herding can include a more radical cultural shift away from wildlife utilization than the necessity of, for example, wage-labour in the hunter-household does.

The discussion in this paper is valid for cultural aspects of animal production generally. I shall therefore not go into any detail on an evaluation of *the different forms* of animal production seen in relation to each other.[7] The following examples from reindeer herding will therefore serve to illustrate some central cultural issues.

The first example deals with *forms of husbandry*. There are naturally many different types of operating systems in reindeer herding taken as a whole, but a more technical discussion of these systems would fall outside the scope of this paper. What interests us in this context is the relationship between "open" and "closed" herding. Open herding is an extensive operational form in which the animals are left to themselves most of the year and are, for example, only rounded up for slaughtering. Closed herding, in contrast, means that the animals are tended intensively year round and that actual breeding tasks such as choosing animals for breeding purposes, etc. take place.

We have many examples on record of how precisely the relationship between closed and open herding in Eskimo reindeer herding has placed people in what we can call, using an expression from psychology, a *double bind situation*: no matter what one does, it results in problems. Closed herding, which most probably yields the greatest profit, is labour intensive and can prevent people from going out hunting and sealing. Open herding offers much better possibilities for hunting and sealing, but consequently one loses, almost totally, control over the reindeer herd. Reindeer herding in Alaska has, for example, swung between these two types of operative systems, causing major problems (see Lantis 1950:32; Naylor et.al. 1980; Stern 1980. See also Seguin 1991:19 ff for a description of the problems in Canada).

If ever reindeer herding in Greenland (and other animal production) is intended as a part of a multifaceted production strategy, one must resolve these basic conflicts between operational methods and cultural value systems. It is in no way sufficient just to attempt to solve the problems using administrative or technical initiatives (corral systems, systems for earmarking, slaughtering facilities, etc.). It is also important to bear in mind that the labour force in reindeer herding in Greenland - both in the Itinnera operation and the Qipisaqqu operation - until now has been *gender-specific* and only involving men. In this way, Greenlandic reindeer herding differs from both household-based sheep breeding in Greenland and pastoral reindeer herding outside Greenland. The gender division amplifies the above-mentioned double bind situation.[8]

The next example deals with *forms of ownership*. Common to most animal production systems is that the system's elements - as a rule, the animals themselves - are *privately owned*. On the basis of historical as well as actual experiences with sheep breeding and reindeer herding in Greenland, we can point out some central issues in relation to the privatization of animal resources. There are especially three problem areas which are important: privatization means *intervention in the food chain*, it results in a *competition for land use*, and it has a tendency to *increase social inequality*.

Privatization means that individual elements in the natural food chain are privatized. This could, for example, be the middle element in the chain plants/land - herbivores - carnivores. The intervention in food chains has partly, important bio-economic implications and partly, there are ecological consequenses in situations where several types of animals principally utilize the same foodbase.

Competition for land arises where the privately owned animal production occupies land areas which are utilized under other forms of exploitation of renewable resources through hunting and freshwater fishing.

Finally privately owned animal production means that individual people have exclusive access to the animals and the products/incomes derived from them. The exclusive use of common resources (land and plants) makes it possible for these individuals, by means of their right of private ownership of the animals, to transform common resources into privately owned products. This situation has socio-economic consequenses which are especially noticed in villages which otherwise rest upon a many-sided utilization of resources.

A readily available alternative to private ownership is the right of *co-*

operative animal ownership. Co-operative Eskimo reindeer herding has been attempted in Alaska without success. One of the major problems here was that the co-operative was not *culturally* sustainable because it could not create and maintain cultural and social status which precisely in Alaska was traditionally connected to a combination of social differentiation and generosity (see, among others, Olson 1969 + 1970; Rainey 1941). In Greenland, reindeer herding in Itinnera has been organized as a co-operative since 1978 under the Kapisilinni Tuttuutileqatigiit. But not even in the more traditionally equalitarian Greenlandic context has the co-operative functioned particularly well (for an in-depth explanation of this, see Rasmussen 1992b). And regardless of whether the form of ownership is co-operative or private, the above mentioned conflicts arise between reindeer herding and other utilization of the land areas. In reindeer herding in Itinnera there are conflicts vis-à-vis caribou hunting, and in reindeer herding in South Greenland there are conflicts vis-à-vis trout fishing and ptarmigan hunting (Rasmussen 1992a).

An increased national meat production in Greenland on the basis of a multifacetted strategy must necessarily find a solution to the unwanted socio-cultural consequences of the said forms of ownership.

My last example deals with the relationship between *sustainability and predictability*. Generally speaking, the possibility of a sustainable utilization and management of the living resources are better, the higher the degree of predictability. But biological and complicated ecological processes run almost chaotically. How can one achieve a sustainable development, if one has great difficulty in predicting - and consequently controlling - the course of events? The traditional hunting and sealing society's way of dealing with this problem was to spread the utilization of resources out over many types. But in a modern animal based economy as in the Greenlandic case, and maybe with increased national self-sufficiency as a goal, such a strategy is much too uncertain. The Euro-American strategy involves eliminating the uncertainty and increasing the predictability through strongly controlled breeding practice and an industrialized animal production which tears the animals away from their natural social and ecological context.

This issue cannot be solved by technical means. It deals with the relationship between culture and the exploitation of nature. Let me give a small example:

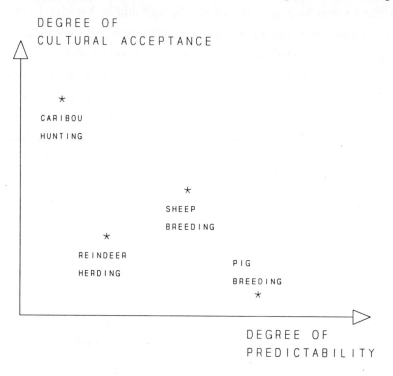

Figure 3

In this simplified example, pig breeding is only included in order to emphasize the point that there are animal production systems with a very high degree of predictability, like for example highly mechanized pig breeding which nevertheless lacks cultural acceptance in Greenland. And there are forms of utilization which are culturally sustainable but almost lacking in possibilities for the control of stock development, etc., as for example caribou hunting.

Conclusion

In this paper I have dealt with a few selected aspects of the complex whole which is called *cultural sustainability* in relation to animal production systems. In this connection I have cursorily cited experiences from Alaska and Canada. I wanted to stress here that a strategy for sustainable utilization of living resources can*not* be narrowed down to including only economic and, in a broad understanding, ecological parametres. If production systems do not have cultural acceptance, they cannot function. It can almost be stated so simply. What is left is the challenge: to identify *which* cultural factors promote or block the introduction of new animal production systems.

Acknowledgements

Field work in the two Greenlandic reindeer herding areas was carried out in 1990 and 1991, and archival research in 1992. I am indebted for guidance and hospitality to the reindeer herders in South Greenland, Ole Kristiansen and Stefan Magnusson, and to the chairman of the reindeer owners co-operative in Kapisillit, Daniel Lukassen and his family. I am likewise indepted to the archivist, Ulla Pagh Andersen, and her colleagues at the Royal Greenland Archive (KNI's Centralarkiv) in Copenhagen for their great readiness to help. Even if it is not made explicit, a substantial part of the archive material forms the basis for discussions in this paper. Research in 1990 and 1991 was supported financially by grants from the Danish Social Science Research Council, and the work in 1992 was supported by a grant from the Commission for Scientific Research in Greenland. The paper was translated into English by Ph.D. David Lipscomb.

Notes

1. In this connection, one does not take into account the economic significance of Denmark's contribution to Greenland's national budget with a capital transfer of approx. 3 billion Danish kroner yearly without a background in Greenlandic production.

2. In Greenland animal exploitation serves as an important cultural point of reference also for people outside the primary production sector. For discussion and documentation of genealogical, socio-economic, and demographic factors' significance for identity-formation, identity-management, and cultural transmission, see Rasmussen 1983; 1986; 1987; 1989.

3. The marine ecosystem around Greenland contains in places concentrations of biomass with high primary productivity in quantities of phytoplankton and zooplankton. These ecosystems are more productive than the terrestrial tundra-biomes. (Productivity is measured here as net annual plant production per areal unit (metric ton per square km), see Stenseth 1991:19-21)

4. During the very first period of reindeer herding in Alaska, Chukchi herders from Siberia were called upon as chief herders and herding instructors. Danish agronomists helped to start Greenlandic sheep breeding in, for example, the establishment of operational and experimental stations in Nuuk and Qaqortoq.

5. In this paper, I shall not take sides in the discussion for and against past analyses of the resource situation and of the human demographic situation in the three countries. Ealier Greenlandic discussions not only mentioned the size of the caribou herds, but the situation in the country's principal profession at that time, seal hunting, was also discussed. The discussions concerning the size of the seal herds were the basis for the decision to introduce experimental fishery and experiments with sheep breeding. The intention was that animal production should first and foremost employ Greenlanders who for one reason or another were not able to hunt seals from a kayak. See, for example, the exposition in Oldendow 1935:299 ff.

6. The growing diversity of animal production systems is not altogether a positive sign in the global context. The ever-increasing threat to biological diversity, and the political-economic threats to *cultural* diversity lead to, among other things, an increased diversity in animal-based production systems. *This form* of increased diversity is, thus, an expression of, or a result of, the increased pressure on the earth's biological and cultural

resources. At the same time, purely structurally and commercially founded mechanisms in, for example, the EEC cause the dismantling of a range of animal- and plant-based production systems!

7.   Nevertheless it is worth mentioning that sheep breeding in contrast to reindeer herding is strongly weighted by a high level of costs (artificial fertilizer, feed, animal shelters, etc.).

8.   It is probably also important in connection with the question of operational methods and cultural value systems to pay attention to the fact that the manager of the reindeer operation in Itinnera as well as the Greenlandic owner of one of the two South Greenlandic reindeer herds both grew up in sheep breeder families (in Neriunaq and Eqaluit, respectively).

References

Anonymous, 1906, Husdyrhold i Grønland. Betænkning afgivet af Udvalget til belysning af mulighederne for Husdyrhold og Rensdyravl i Grønland. *Det grønlandske Selskabs Aarsskrift* 42-66.

Anonymous, 1946, 1947, 1948, *De Forenede Grønlandske Landsråds Forhandlinger.*

Anonymous, 1981, *Betænkning om Den erhvervspolitiske udvikling i Grønland.* 1. Del. Landsstyreområdet for erhvervsmæssige anliggender, Greenland Home Rule, Nuuk

Anonymous, 1989, Cost-Benefit analyse af fåreavlen i Grønland. Delrapport til *Betænkning fra Fåreavlererhvervets arbejdsgruppe.* Direktoratet for Bygder, Yderdistrikter og Landbrug. Greenland Home Rule, Nuuk.

Anonymous, 1990, *Betænkning fra Fåreavlererhvervets arbejdsgruppe.* Direktoratet for Bygder, Yderdistrikter og Landbrug, Greenland Home Rule, Nuuk.

Beach, Hugh, 1984, "Developing Spirit in Northwest Alaska". *Ethnos* (3-4): 285-304.

Beach, Hugh, 1985, "The Reindeer-Caribou Conflict in the NANA Region of Alaska: A Case Study for Native Minority Rights Issues". *Nomadic Peoples* 15 (February): 1-21.

Beach, Hugh, 1986, The Saami in Alaska: Etnic Relations and Reindeer Herding. *Contributions to Circumpolar Studies.* Uppsala, The Department of Cultural Anthropology, Uppsala University. 1-81.

Dahl, Jens, 1989, "The Integrative and Cultural Role of Hunting and Subsistence in Greenland." *Études/Inuit/Studies* 13(1): 23-42.

Duerden, Frank, 1992, "A Critical Look at Sustainable Development in the Canadian North". *Arctic* 45(3): 219-225

Lantis, Margaret, 1950, "The Reindeer Industry in Alaska". *Arctic* 3(1): 27-44

Naylor, Larry and Richard O. Stern et al., 1980, "Socioeconomic Evaluation of Reindeer Herding in Northwestern Alaska". *Arctic* 33(2): 246-272.

North, Dick, 1991, *Arctic Exodus. The Last Great Trail Drive*. Toronto, Macmillan of Canada.

Oldendow, Knud, 1935, Naturfredning i Grønland. *Det grønlandske Selskabs Aarsskrift*. 9: 299-314.

Olson, Dean F., 1969, "Territory, Village identity, and the Modern Eskimo Reindeer Manager." *Canad.Rev. Soc. & Anth.* 6(4): 248-257.

Olson, D.F., 1970, "Cooperative Ownership Experiences of Alaskan Eskimo Reindeer Herders." *Human Organization* 29(1): 57-62.

Postell, Alice, 1990, *Where Did the Reindeer Come From? Alaska Experience, the First Fifty Years*. Portland, Amaknak Press.

Rainey, Froelich, 1941, "Native Economy and Survival in Arctic Alaska". *Applied Anthropology* 1(1): 9-14.

Rasmussen, Knud, 1907, "Rapport til Indenrigsministeriet over Renbejte-Undersøgelses-Ekspeditionens Rejse i Grønland, Sommeren 1905." *Atlanten* (2): 43-58.

Rasmussen, Hans-Erik, 1983, *Social endogami og symbolbrug i Vestgrønland*. Vol. 1-3. Department of Eskimology, University of Copenhagen.

Rasmussen, Hans-Erik, 1986, "Genealogier og social stratifikation. Eksempler fra Grønland". *Fortid og Nutid* (2): 95-110.

Rasmussen, Hans-Erik, 1987, "On Socio-Genealogies in South West Greenland, 1750-1950. Some explanatory sketches". *Acta Borealia Oslo* (1-2):43-52.

Rasmussen, Hans-Erik, 1989, *The Moravian Congregation at Lichtenfels, South West Greenland, 1758-1900*. A Demographically and Genealogically Survey of a Closed Population. Department of Eskimology, University of Copenhagen.

Rasmussen, Hans-Erik, 1992, "Den sydgrønlandske rendrift". *Grønland* (1): 7-26.

Rasmussen, Hans-Erik, 1992, *Reindeer Management in Greenland: Cultural and Economic Problems*. The First Nordic Arctic Research Forum Symposium, Gilleleje, Denmark,

Seguin, Gilles, 1991, "Reindeer for the Inuit: the Canadian Reindeer Project, 1929-1960". *Muck-Ox* (38): 6-26.

Stenseth, N. Chr., 1991, Forvaltning av biologiske fellesressurser i et lokalt og globalt perspektiv. *Forvaltning av våre fellesressurser. Finnmarksvidda og Barentshavet i et lokalt og globalt perspektiv (pp 11-70)*. Oslo, Ad Notam Forlag.

Stern, Richard O., 1978, *Alaskan Eskimo Reindeer Herding and Value Systems*. The Subsistence Lifestile in Alaska Now and in the Future, The School of Agriculture and Land Resources Management, University of Alaska Fairbanks,

Stern, Richard Olav, 1980, *I used to have lots of Reindeers*. The Ethnohistory and Cultural Ecology of Reindeer Herding in Northwest Alaska, The State University of New York at Binghamton.

The Northern Québec Isolated Communities Database: Structure, Issues, Challenges

Pierre Saint-Laurent

Introduction

Hydro-Québec is the public utility serving electric energy needs in Québec, Canada. It is a regulated (quasi-) monopoly, as most electricity produced and distributed in Québec (75.3 % is produced, and 90.3 % is distributed, by Hydro-Québec, according to the 1989 edition of Le Québec Statistique) is under its direct responsibility. Most of Hydro-Québec electricity generation is of a hydroelectric nature (93.2 % according to the 1991 Hydro-Québec Fact Sheet), and may involve large dam projects in the northern parts of Québec. Additional electricity is produced by a small nuclear generating plant (Hydro-Québec has only one such generating faciliy, called Gentilly-2), as well as thermal equipment in remote locations where connection to the main power grid is unacceptably costly or complex. Thermal power is also used to respond to peak load needs, which represent an important consideration in northern latitudes (with large yearly temperature variations) such as those under which Québec is located.

Hydro-Québec is constructing an extensive database which will contain large amounts and types of data pertaining to isolated communities located in northern Québec, roughly north of the 48th parallel. This database, when completed, will contain the most extensive array of data on the isolated communities of northern Québec that it is possible to gather and process. As such, it will contain data of a demographic, political, social, anthropological, economic, health and family nature (this list is not exhaustive). It will be made available to researchers that may have a need for data specific to this region of the world. Foreign researchers will be welcome to access it, whether for comparative studies purposes or for studying northern Québec in and by itself. The GÉTIC (Groupe d'ÉTudes Inuit et Circumpolaires) of Laval University is the major collaborator in this project with respect to methodological questions.

There are many reasons for which this database is being set up:

1) for internal purposes, i.e. for Hydro-Québec's need to know and understand, in a retrospective as well as an ongoing fashion, the many aspects, perspectives and constraints of life in northern regions where hydroelectric projects may be or are planned or located, as well as address socio-economic questions relevant to such projects, such as questions of economic development, of incomes, of health status, and the like;

2) for external purposes, i.e. to make accessible and available to outside researchers a large quantity of data which would otherwise be hard to access and obtain by individual effort;

3) for "corporate citizenship" reasons, i.e. to play an active leadership role among the different organizations that use and/or collect data on isolated communities in the northern regions of Québec. Indeed, one of the premises underlying this effort is the advantage gained by all those participating in it. On one hand, government agencies and other organizations that have social responsibilities in northern Québec- there are many- as well as researchers and other participants, collect data for their own purposes. The construction of the database is predicated on the obtaining of data from these entities. On the other hand, these same data-collecting entities stand to gain by participating in Hydro-Québec's effort, at least in two ways: first, they will supply data to the database, in exchange of which they will be contributing to -indeed, making possible- a simple way of getting the broadest view possible of the northern Québec situation. Second, they will be inputing their data in a system for which data comparability, homogeneity, stability in time, and the like, are considerations of paramount importance. Hence, this quality control, as it were, may contribute to the production of better, more homogeneous, more comparable, data than otherwise economically feasible.

The following text is divided in three parts. The first part will present a rough outline of the structure of the database. The second part will expound some of the issues surrounding the construction of the database, while the third and last part will represent in fact an invitation to interested researchers to share ideas, this invitation being focused on certain specific ideas and thoughts. Finally, a conclusion will summarize the text itself.

Structure of the Database

The database is basically structured so as to exclude no data on *a priori* grounds. Indeed, great care is taken in the preparation of the structure to ensure that from a methodological point of view, no exclusions can be due to construction bias or researcher preference. This does not mean that difficult choices do not have to be made: more on this later.

There are 13 categories of data, corresponding to a "social sciences logic", so to speak: these categories, at least in our view, represent "natural" categories, i.e. that would come to mind more or less spontaneously to a social sciences researcher if asked to name categories meaningful to his work, or at least that in his opinion, would best serve the broad, general needs of the social sciences in a Nordic/Arctic perspective. These categories are the following: demographics, territory structure, housing and municipal services, government expenditures, education, health, manpower and employment, wildlife use, transport and communications, income, justice and law enforcement, electoral and political participation, and adjustment factors (these factors are statistics that allow adjustment of the data contained in other categories so as to permit comparison with data from other regions of Québec, Canada, and eventually, other regions of the world). These categories are the product of extensive work performed by members of the GÉTIC, mentioned above.

These categories are quite compatible -indeed, they are mimicked from usual "meridional" categories used to structure data arrays (In this area, one can refer to, e.g. United Nations (1970). An Integrated System of Demographic, Manpower and Social Statistics and its Links with the System of National Accounts, New York, U.N.; Statistics Canada (1980). Standard Classification of Industries 1980. Ottawa, Supplies and Services Canada.) Moreover, their very broad scope allows inclusion of a maximum number of observations without resorting to data selection and/or manipulation. It is indeed the objective to maintain the lowest level of aggregation possible, leaving the aggregation level decision itself to the researcher. However, there are problems that appear at other levels in the construction of the database. These problems will be addressed shortly.

At a greater level of detail, certain subcategories contained in the database are presented in the list below. As one can see, there is very little surprising or new about such classifications. And indeed there should be no surprise in a database: the objective is at all times to secure maximum comparability and compatibility with other arrays of data available elsewhere, and with which there may be joint use of observations.

Exhibit 1
Description of Selected Data Contained in the Northern Québec Isolated Communities Database

Demographics: population by 5-year age groups, births, deaths, migration, matrimonial status by 5-year age groups.

Territory structure: area of villages, administrative divisions, administratively determined categories of territories; characteristics of territories.

Housing and municipal services:
Housing: housing by number of occupants, by ownership/rental status, by type, by age of owner/occupant, by amount of rent.

Municipal services: snow removal, fire protection, garbage disposal, drinking water, sewer, municipal office, municipal vehicle expenditures and descriptive data; fiscal revenues, taxes by type, property values, energy consumption by user, telephone use, television and cable use.

Government expenditures: Federal and provincial expenditures (total and per capita), by program, by department, by community.

Education: Number of students by 2-year age groups, number of students by school year (scholastic level), by sex; number of university students, by sex; number of schools, by type; number of students by language of schooling; number of teaching personnel by scholastic level, by sex, by school status (private vs. public/government); non-teaching personnel by type of school; number of teachers, of courses and participation rates in continuing education; costs and revenues by type of school.

Health: Number of hospital patients by 5-year age groups, by sex; number of diagnoses by diagnosis type; number of hospital admittances and hospital days by diagnosis type; number of evacuations by destination; number of evacuations by diagnosis type; number of work-related accidents by occupation type; number of professional illnesses by occupation type; number of persons eligible to the universal health care system by 5-year age groups, by sex; number of patients by institution, number of professional visits, number of dentist visits by type; number of personnel by type; costs by type of institution.

Manpower and employment: Labor force by sex, unemployment rate by sex, number of self-employed workers by sex, number of workers by employment sector by sex, full-time and part-time workers by broad employment sector.

Wildlife use: hunting ground area/size, by type and category; participants in the Income Maintenance Program (The Income Maintenance Program, negociated with the Cree in 1975 in the context of the James Bay-Northern Québec Agreement, under which traditional activities (specifically, hunting and trapping) are subsidized on a per-day basis, irrespective of sex. This source represents a major percentage of total income for its participants. For more information, see the text of the Agreement), by sex; number of days hunting; edible weight by type of game, per capita; revenues from fur trapping; average prices of furs; costs related to hunting and trapping; prices of meat, by type of game; market value of meat and furs, by broad category.

Transport and communications:
Transport: number of trips, by mode; weight and value of merchandise carried; freight cost/Kg; number of departures and arrivals/year; number of aircraft, boats, trucks, cars, all-terrain vehicles, snowmobiles, buses and taxis.

Communications: number of telephone connections; number of radio and television stations, by type; subsidies to radio and television broadcasting, by source; number of programs/week, by type; languages used in broadcasting.

Income: wages by sex, sector of occupation (public, private) and economic sector (primary, secondary, tertiary, arts and crafts, commerce, industry, local, aboriginal, external firms); welfare recipients, bt program; personal income per capita; cost of living adjustment; disposable personal income; standardized disposable personal income; income by $5000 bracket, by sex; family income by $5000 bracket; household income by $5000 bracket; average and median incomes by community.

Justice and law enforcement: number of police personnel; number of personnel by ethnic origin; crimes by type, by 5-year age group, by sex; number of crimes by type; costs of judicial services; number of court personnel by type; number of days of penalties served, by type; number of prison sentences by length, by type of institution; detention rates by ethnic group; crime rates by 5-year age group.

Electoral and political participation: voting-age population by political level (local, provincial, federal) ; registered voters by political level; number of voters and abstentions, by political level; participation rates, by level and sex; voting rates for elected governments; voting rates by political party.

Adjustment factors: CPI and inflation rate, Québec and Canada; average income tax rate; expenditure rates on housing, food, clothing, transport, leisure; local, regional, provincial, national expenditure rates; average total income, Québec and Canada; average disposable income, Québec and Canada; savings rate, local, Québec and Canada.

Issues in Construction of the Database

Exhaustivity: It should now be obvious to anyone having perused exhibit 1 above, and with any familiarity with Nordic/Arctic situations, that the agenda consisting in filling in the empty boxes, so to speak, represented by the data categories, is a more than ambitious one. Indeed, it should be apparent, and expected, that only a fraction of those categories can be documented with any form of data, let alone reliable, verifiable ones. Thus, it should be understood that the northern Québec isolated communities database has a very definite "ideal" aspect to it, that is, the initial construction of the categories must include those that researchers would like, in an absolute sense, to see included in such a data collection and management effort. Once again, we are back to considerations of bias avoidance: such an approach ensures that those "empty boxes" lie there waiting for data to eventually be included in them, when better data collecting facilities and techniques, better modes of communicating the data, and user demand, make the effort of filling them worthwhile. By including a maximum number of data categories, irrespective of whether there actually are data to be categorized thus, minimizes the bias produced by selecting categories on the sole criterion of contemporaneous existence of observations, even though the database is a permanent structure which will exist for an indeterminate, theoretically infinite, amount of time.

There are two other, very compelling, reasons to create an "ideal" structure (from a methodological standpoint) from the outset: first, the structure itself may become a framework, useful to data collectors and gatherers not yet

109

involved in the database, to focus their efforts around an array which a) is compatible with "meridional" data and data from other regions of the world, inasmuch as such compatibility is feasible; and b) already contains time-series and cross-section data which can always be refined, completed or complemented by further observations. Secondly, this "ideal" structure is directly function of databases already in existence, mostly meridional. The categories included in the northern Québec isolated communities database are nothing new: on the contrary, they mirror equivalent categories related to data-collecting efforts in the southern parts of Canada. The issue in this case is not so much to theoretically categorize ideal data that may some day become available; it is rather to make available data comparable with similar categories in other regions, on one hand, and on the other to allow researchers free rein with respect to the actual data-collecting effort (which is in itself the most formidable challenge of this whole endeavor), without having to grapple with hard methodological questions or to resort to makeshift techniques in dealing with comparability of their data with the observations already accumulated by others.

Participation of data-collecting entities: The collecting of data remains under the full control of the agencies and researchers in the field. The structure of the database may suggest ways in which to go about this task, but in no way does the construction and updating of the database bear any mandatory or coercitive aspect. Quite the contrary, the cooperation of a large number of otherwise independent agencies and individuals is necessary if this database is to become operational. In other words, this is fully a cooperative effort, which means that all parties involved must look forward to some benefit in participating in the endeavor. At this point, positive reactions and participation have been forthcoming from all those involved up to now. However, the issue remains, as in all collaborative efforts, as to the permanence of commitments, that is, entity agendas may change in time to the point that the supply of data to the database from certain sources may be changed, reduced or even jeopardized. This is akin to saying that the database will be the sum total of the efforts expended by external entities -that is, external to the control of those maintaining the database- to collect data pertaining to the northern parts of Québec. However, if present participation and curiosity is any indication, it seems that economies of scale and data validation prospects are such as to generate long-term participation.

International interest: One of the main objectives of this database project, and one of the wishes of its promoters, is to make available to a wide international audience of researchers and agencies a homogeneous, standardized, useable array of observations which may help them in their own research, social, funding, regulatory, administrative, or other functions. To use a stovetop analogy, the proof of the pudding is in the eating. This means that ultimately, it is the degree to which this database is used by entities other than Hydro-Québec and other Québec agencies and scientists, that will be the clearest indication of success with respect to this issue. It should be stressed at this point, however, that input from the international community would indeed prove most helpful in the earliest stages of database elaboration. In the long run, this database should be part of a worldwide effort in understanding the parameters and specificities of Nordic/Arctic regions in a manner that transcends national boundaries so as to focus on ethnic, economic, social, language, and development issues, to name but a few. Hence, there is definite and ample room for input, especially at this early stage of development and construction, from many members of the international Nordic/Arctic research communities, the residents of these areas, and the agencies with responsibilities in these regions. After all, these entities stand to gain in participating in this effort, as their own work will eventually and/or ultimately be enhanced by initiatives of the sort expounded herein.

Methodologies: Implicitly, the methodology adopted for the northern Québec isolated communities database is the one developed by Statistics Canada for its data collection duties across Canada, and augmented/modified by the research efforts of the GÉTIC. This raises the question of international comparability/compatibility: Can we be sure that these data can be combined with data pertaining to other countries? *Prima Facie*, there seems to be no apparent problem, as a statistical methodology such as the one developed and used by Statistics Canada clearly is normalized in an international fashion: When Gross Domestic Product (GDP), or infant mortality rates, are measured, there is direct comparability of these numbers internationally (at least across industrialized, developed, "Westernized" countries). Are we thus out of the methodological woods, so to speak? Not necessarily.

There are (at least) two reasons why one can, and should, keep a watchful eye on potential methodological problems developing.

First, there may be problems between northern and meridional data within Québec, or for that matter, between northern and meridional data within any country with Nordic/Arctic regions. This could occur if, from a structural point of view, the economies and societies of the two regions are sufficiently different that a unified methodology captures significantly different realities. This raises the question of how to calibrate such differences within a single methodology and also how to reconcile differences so as to restore a unified descriptive body of data. In other words, once it is determined that there exist significant discrepancies (if applicable) between observations in the database and observations to which they are destined to be compared, what adjustment factors can one create to ensure consistency, at least at the national level? The answer, at least in the mind of this author, is not open to simple solutions.

One example of this potential problem should make the point clear. It is well known that self-consumption phenomena (the fact of spending time and effort in harvesting wildlife and other resources for one's own consumption) represent substantial proportions of the activities of Aboriginal populations in Nordic/Arctic settings.

One key characteristic of such economic activities is that they are not monetized, i.e. they do not figure in the official national accounts systems as figures expressed in the national currency (the meridional equivalent, albeit in a lesser proportion of total household income, is self-consumption activities by farmers). This means that, since market forces do not put values on substantial portions of the economic activities of populations, the only way to obtain statistics documenting them is by constructing an imputed value.

Such an imputation of values must then be based on , e.g., market values for the same products in other regions (or eventually in other countries, with proper adjustments). However, how should one proceed if there are no true markets for such products, as would be the case for most of Nordic/Arctic self-consumed products except perhaps furs? Should one impute the price of, say, beef to self-consumed caribou meat, under an alternative cost approach based on what could potentially be consumed instead of the self-harvested and consumed product? What about the cultural values underlying the fact of harvesting one's subsistence? What about the structure of the market for caribou meat that could theoretically exist, and that the researcher is implicitly trying to mimic in its price-generating aspects? As one can appreciate, this represents an area of deep and difficult questions.

Second, a related problem raised by such an issue is the non-comparability that this may create between northern data pertaining to different countries. Indeed, if makeshift methods are adopted, on a case-by-case basis, to try to make compatible northern and meridional data within each country where such data-management efforts are undertaken, what guarantee is there that northern data will then be comparable from one country to another? In other words, if there are as many solutions to the north-south data problem, as it were, as there are different databases, how can one then take international data and be sure that the imputations, to take the example just given, have been performed in an equivalent fashion? This suggests that if there truly exists an (implicit) international agenda to further understanding of Nordic/Arctic questions via gathering of high-quality, consistent, interregionally and internationally comparable data, then this is far from a moot point.

Aggregation level and confidentiality: Another important issue stems from the small population sizes in Nordic/Arctic regions. The combination of very small community size and high level of detail in the data categories illustrated in exhibit 1 above will lead to situations where a data category contains but one observation -making it possible to trace back and identify the particular individual to which the observation corresponds. Such nominative data are unacceptable in a publicly accessible database. However, this will raise important methodological questions as to the correspondence of the data categories, created in a meridional context in which large populations allow to circumvent such confidentiality problems except in rare instances. In other words, data-gathering entities will supply the database with the lowest level of aggregation possible which does not jeopardize confidentiality. There is no guarantee, however, that these data meaningfully correspond to one of the categories already created. Also, the resulting level of aggregation may be so high as to preclude interesting research. As a purely hypothetical example, if a researcher wants to study the crime rate of White women between the ages of 15 and 19 by community in northern Québec, he may possibly find out that only, e.g., the crime rate of all women by community, or the crime rate of women aged between 15 and 19 for all of northern Québec, is made available to him/her.

The Challenges Ahead

The preceding discussion has raised some (not all, by far) important issues in the construction of the northern Québec isolated communities database. This section will represent in fact an invitation to interested researchers to interact with this author and others interested in this area, to further these issues as to make available to a broad international body of users of statistics, the best quality and quantity of observations possible, as well as improved methodologies to ensure compatibility and validity of efforts.

Exchange of ideas: It is obvious, to this author at least, that a wide constituency of Nordic/Arctic researchers and agencies are potentially interested in the elements expounded and the issues raised herein. In this perspective, it seems important that those scientists and researchers that can contribute to this endeavor, do just that- come forward and participate in the exchange of ideas, comments, and suggestions that any true research effort is meant to elicit. Many stand to gain in so doing, and in this way, one of the objectives of this Symposium -to bring together professionals involved in Nordic/Arctic regions to increase and improve research, understanding, and sustainable development- may be furthered.

Methodological section within the NARF Symposium: This author submits that, if interest justifies such action, a section devoted to data and methodological questions could be considered for annual discussion in the context of the Nordic Arctic Research Forum Symposium. Indeed, this would represent an excellent context to share specific research, update knowledge and possibly establish data-collecting and methodological joint ventures.

NARF methodological committee on normalization issues: If need be, NARF could very well represent the right context to establish an ongoing committee in charge of collecting current research on methodological issues of the type raised in this paper, suggesting avenues of normalization and participating, if needed or desirable, in the implementation of normalized methodological protocols.

Conclusion

This paper has presented efforts made by Hydro-Québec to produce an extensive database on isolated communities in northern Québec. It was seen that data categories are elaborated from an "ideal" standpoint, so as to avoid bias in category selection on the basis of contemporaneous availability of data.

Some of the issues raised by this endeavor are the following: exhaustivity of the data categories created, participation and long-term commitment of data-gathernig entities, and international interest.

Another important issue raised is the matter of methodologies. The methodology retained for construction is essentially the one used by Statistics Canada for its nationwide data-collecting duties, as adapted by the research efforts of the GÉTIC. Certain issues were raised as to potential

115

problems that could occur in this context: first, north-south data compatibility, particularly in light of the imputation of values that is necessary in Nordic/Arctic contexts; second, compatibility between Nordic/Arctic data from different countries. These methodological questions raise hard questions of an international nature -and hopefully, of international interest.

Finally, certain forthcoming challenges were presented to motivated readers, in the spirit of international cooperation on these issues. Certain mechanisms were proposed to create permanence and critical mass with respect to potential widespread research and applied work in these areas.

References

Annuaire d'Hydro-Québec (*Hydro-Québec Yearbook*), 1989, Hydro-Québec, Montréal.

Charest, Paul (under the direction of), 1992, *Projet de banque statistique Métrinord - Rapport de la phase I:* étude de faisabilité. GÉTIC, université Laval, Sainte-Foy (Québec).

Duhaime, Gérard, 1987, *Le pays des Inuit: la situation économique 1983.* Coll. Économie politique du Québec arctique, rapport de recherche no 3, département de sociologie, université Laval, Sainte-Foy (Québec).

Égré, D. and P. Senécal, 1992, Human Impacts of Hydro Projects in Remote Areas, *Communication, 12th annual meeting of the International Association for Impact Assessment* (IAIA), Washington, Aug.19-22, 1992. Hydro-Québec, Montréal.

Hydro-Québec Fact Sheet, 1991, Montréal, Hydro-Québec.

James Bay-Northern Québec Agreement, 1987, Éditeur officiel du Québec, Québec.

Le Québec Statistique, 1989, *Les publications du Québec*, Québec.

Statistics Canada, 1980, *Standard Classification of Industries 1980.* Ottawa, Supplies and Services Canada.

United Nations, 1970, *An Integrated System of Demographic, Manpower and Social Statistics and its Links with the System of National Accounts.* New York, U.N.

Trends in Development and Research in the Arctic Regions of Russia

Vladimir I. Pavlenko

Introduction

The aspects under consideration in this article cover most current questions of contemporary development of Russian Arctic : economic, social, ecological. There are priorities of Russian science in Arctic exploration and structure of research programs that were revealed in the article. Special feature of modern period of time consists in drastic changes of economic mechanism, reduction of centralized management and funding of science. Directions of international scientific cooperation are proposed.

The Russian Republic has an Arctic zone of 2.5 million square kilometres. The development of Arctic research in our country has a long history, dating back to the 18th century. The Arctic regions and seas always played an important role in the territorial division of labor, mainly due to the highly efficient natural resources (hydrocarbon raw materials, non-ferrous metals, gold and diamonds) and valuable bio-resources. The richest reserves of oil and gas in Russia are found in the northern and Arctic territories. These regions possess the greatest amounts of known reserves of nickel, copper, tin, gold, diamonds and other minerals. The Arctic zone occupies a special place in the defense strategy of the country.

Economic and social development

During the last three years we have had an unusual situation in the research and development of our Arctic and northern regions. The transition to a market economy in the Russian northern and arctic regions is marked by the aggravation of economic and social problems. In comparison with the other areas, the economic situation in the northern and Arctic regions is becoming strained much faster. The volumes of mining production - the main sector of the economy in these regions - is continuing to decrease. For example, during the last three years oil production only in the Arctic regions of West Siberia fell by 30 %. Production of various metals, including gold, decreased as well. This process has been accompanied by decreased labor productivity (by 34-42 % in various industries) and intensified industrial construction.

Construction rates are slackening too. In comparison with the period of 1982-1986, the volume of construction in 1987-1991 was only 55 %. The volume of housing construction, especially in the arctic regions of Siberia and the Far East, sharply slackened.

The energy complex and transportation system have become less efficient. The rise in energy and transport prices has not resulted in adequate improvement in working condition in the enterprises of these industries. Energy construction was stopped last year in a majority of the Arctic and northern regions. The shipment of goods - including food and equipment - has constantly fallen during the last three years.

The falling behind of the economy has led to growing social tension in the northern and Arctic regions. During the last five years the living standards dropped by more than 2.5-3 times and more. The relative attractiveness of work in the North has fallen.

With the economic recession, there are fewer jobs and many people have departed from the region, worsening the situation in the various industries and in the social sphere. The reduction in the labour force has created a new problem for the Arctic - a problem of conservation of abandoned houses and the "closing" of some settlements (especially at Chukotka). These processes are mostly the result of the disintegration of the USSR and some political and economic decisions taken by new states (Ukraine, Azerbaijan, Byelorussia).

Small businesses are developing very slowly and do not really influence the economic development of the Arctic. Now in the Russian north there are about 600 small business enterprises. Considering the backwardness of the traditional sections of the economy of the native Arctic peoples and their low living standards, we can forecast the further worsening of the social and economic situation of the northern native peoples. The current federal government policy does not take into account the specific character of Northern and Arctic development.

Political and social activity among aboriginal peoples is growing. They have formed a lot of national associations. Each region in the Arctic and the North has its own association. These organizations were established by the native peoples in their struggle for national interests, culture, language, environment, and the development of traditional industries.

The socio-economic situation in Russian Arctic and northern regions in the last few years gives rise a lot of new problems. What will be the directions for developing industry and social infrastructure under market economy conditions ? What are the areas for developing small business ? Who will pay the high cost of social services ? Who and how will cover the ever increasing expenses for Arctic products and how?

The natural desire on the part of people to run their own affairs is not limited to aboriginal people. Also people of different nationalities descending from other regions of the former USSR (Ukraine, Byelorussia, Azerbaijan) support these activities. Arctic and northern land is now their land as well. And they want to struggle for freedom from control by the central government.

Self-government means more than acting as an agent in the delivery of services for the federal or provincial government. In Russia there are at present three main reasons for self-government :

- First, there is the natural desire of human beings to run their own affairs;

- Second, aboriginal people believe that they can manage their affairs more successfully than bureaucrats and politicans from Moscow or other cities.

The most important affairs are the ecological problems, development of traditional industry;

- And third, they feel that this process is essential for the future of the next generation.

The ecological situation in the North is getting worse. Scientists and specialists are seeking solutions to the problems of radioactive materials present in the Kola Peninsula, Novaya Zemlya Island and other regions (Norislk, Chukotka, Yamal).

The central government does not relate the development problems of the North to the development problems of the State, or to the tranformation of economic and social patterns while market economy is implemented.

The social, economic and political problems arose rather a long time ago and they are accumulating. The international scientific community wasn't aware of them earlier, but now these problems have become clear. We are trying to find different roads and methods to resolve them. We are trying to learn from international experiences and to get international support. We think that in other states the majority of scientists understands our problems, and that ecological problems do not have national boundaries.

The research and development

We have created a special program to study the economic and social problems for all of Russia, and within this program there is a focus on the Arctic and northern regions. The designing of economic and social development programs from the statewide level to the regional level allows us to ensure a balance of interests at the federal and local levels, to bring together the interests of different industries and regions, and to develop an adequate policy for social affairs as well as science and industry.

Because of the Arctic's rich natural resources, the world community is especially interested in this region. The Arctic is the richest source of fuel, energy, mineral and unique ecological and biological systems in the northern hemisphere (maybe even on the whole planet). The Arctic may even be one of the last sources available, but we should always have in mind that the utilization of its natural resources demands enormous financial expenditures.

The importance of these regions with their not yet exhausted natural resources will obviously increase. The Arctic is also of great importance as a relatively incompletely explored region, which strongly influences the climate and weather of the whole planet. A lot of unexplained phenomena are connected with the atmosphere, hydrosphere and lithosphere of the Arctic.

The research and development of Arctic and northern regions require great financial, material and intellectual resources. Now we are considering these regions as a good field for international cooperation in both science and industry. It gives us all an opportunity to gather our scientific and financial resources and to exchange knowledge and technology.

The analysis of international experience shows that the study of the Arctic is mainly determined by two factors : by the demand of the society to go deep into the new processes and phenomena in these regions, and by the level of science itself, its methodology, methods and equipment. These factors influenced the formation and organization of Arctic research in our country. Incontestable priority belonged to the fundamental research of global processes in various spheres of the Earth : litosphere, hydrosphere, cryosphere, atmosphere, magnitosphere and ionosphere.

In recent years the attention to concretely applied research directly connected with industrial exploration of the Arctic and human activity in extreme conditions has increased very much. I mean the utilization of modern technology in different areas with a view to increased labor productivity (particularly in mining), to maintain the ecological balance, to decrease the negative effects of humans and technology on nature and to improve the living standards of the population.

These problems are the focus of newly formed research centers within the system of the Academy of Sciences. In recent years some new institutes were organized, within the framework of traditional research centers. These institutes are dealing with specific regional, ecological, economic, social, and technological problems connected with Arctic development. Several new centers were created in the northern part of European Russia, in Siberia and in the Far East. There are about 2000 specialists now studying the Arctic.

In 1990 the Interagency Commission on Arctic Research under the Presidium of the USSR Academy of Sciences was created. In 1992 the Joint Scientific Council of Arctic and Antarctic Research was established. The main task of this Council is to coordinate scientific research in the Arctic, to consolidate the efforts of scholars and establish priorities, to provide a common methodological approach to the elaboration of state research programs and to increase the efficiency of international scientific cooperation in the Arctic.

Taking into account the necessity of permanent cooperation between institutes from different regions of Russia, the Arctic Research Center, within the Russian Academy of Sciences, was formed in 1991. The Center is a basic scientific, research and coordinating body of federal, republic and academic Arctic research and development. Our scientific and practical interests cover almost all areas of activity in the Arctic and Subarctic, including the problems of international security and law, sociology and economy, ecology and biology, geology and physics of the cryosphere, medicine, ethnography and engineering.

Practically all research is done in the framework of federal, republic, academic, regional and international programs. In 1991 a joint interagency program called "The Arctic" was started. This program, in which 140 different research bodies take part, covers the priority directions in fundamental and applied sciences.

The proportion of centralized sources is going down, and on the contrary the share of decentralized financing based on concrete contracts is increasing. This fact opens new possibilities for cooperation on specific projects with scientists and specialists from foreign countries on a contract basis.

The democratization of our society, the deepening of glasnost and openness, and the conversion of the defense industry open new possibilities for cooperation in different fields. We are ready to use the infrastructure of Barentsbourgh and Pyramida on Spitsbergen in the most effective way and on an international basis for the promotion of the Arctic research in this region. We are ready to open our scientific bases and research ships as well to international expeditions.

The organization of the research in the areas mentioned above is possible on the basis of scientific research institutes which have permanent field centers in the Arctic. The institutes can also use satellite, rocket, balloon and land observations.

Of particular interest is the creation of special models showing the changes in the Arctic environment brought about by human presence and the influence of these changes on the global processes. It is very important to work out a method to evaluate the consequences of big industrial centers functioning in the Arctic and the frontier regions. The accumulation of human materials in the Arctic can cause irreversible changes in the atmosphere and ocean surface, can disturb the thermodynamic balance and pressure fields. It is necessary to use the information from ground and space observatories because of the complex character of the interaction among the Arctic environment, human substances and external natural factors of space and lithosphere origin. The conversion of the defense industry gives us a real opportunity to participate in the organization of ground, sea and space monitoring. It's very important to study the evolution of Arctic ecosystems under the growing pressures, both human and technological, the influence of natural and human ecosystems on human beings. While developing the Arctic and northern regions it is very important to minimize the negative effects of human activities on wild nature and valuable species of fauna and flora. The IASC determined environmental issues as a top priority.

Our priorities include traditional knowledge as well. The problems of human adaptation and human ecology form part of a special program, "Care of public health in the Arctic and northern regions."

All of these problems were united in the national "Arctic" program. In the last fiscal year we spent 50 million rubles on this program.

In 1991 some institutes from the Russian Academy began research projects with Norway (Barents Sea oil and gas exploration, geophysical problems), Sweden ("Tundra" project), the USA (Beringia), Canada (geological) and others.

We hope that the "Arctic" program will give us the possibilities of combining our national interests with the international scientific community in Arctic research and development.

Conclusion

Despite certain economic difficulties, development of Arctic research in Russia is based on powerful scientific and technical potentials and takes place in the frames of the national program. Approaches, proclaimed in Murmansk, to study the Arctic regions on the basis of joint efforts of different countries are implemented by Russia in bilateral and multilateral projects. These approaches will establish a good and reliable foundation for development of scientific contacts in the Arctic regions.

References

Dictionary of economic and social development of the Russia in 1991, Moscow 1992.

Pavlenko, V., 1992, *Strategy development and research of the Arctic regions*, Moscow.

Sami Bill up in Smoke
- Discourse Strategies within
Sami Ethnopolitics in Sweden

Eivind Torp

On 22 October something quite rare happened in Sweden, the national television news actually featured an item about the Sami! "The Sami are on the warpath" was the headline, so naturally viewers paid extra attention. Later, the news item was introduced in roughly the following words:

"The Government Bill on the future rights of the Sami was sharply criticised when it was discussed at the Sami National Congress last weekend..."[1]

The report that followed was very brief. Viewers saw a person standing by the rostrum, describing the Government Bill in the following way:

"This is a good proposition if you want evidence of how Sweden treats the Sami, but it is completely worthless as basis of any action to strengthen our rights. The only sensible thing to do with it is this..."[2]

After this declaration the speaker left the rostrum, went over to a window, opened it and set fire to the Bill. The audience was completely silent. The only sound the viewers could hear was the clicking of dozens of cameras; the photographers had clearly stationed themselves by the window before the speech was completed.

TV-viewers who watched this news item witnessed something that was quite new in the Swedish context. Reports of any sort on Sami culture are most unusual on national television, and when there is an item, it is almost always an exotic report on reindeer herding. This, in contrast, was a report about the current political situation - and, most unusually, Sami protest actions were displayed so openly. That was something very few people had ever seen on a Swedish newscast before!

If it is so unusual to be confronted with Sami protest actions or demonstrations in Sweden, what was the background to the Sami being "on the warpath?"

Background

For the last 15 years, the question of Sami rights has been the object of official enquiries in Finland, Norway and Sweden.

Regarding Finland, a Sami parliament has existed since the middle of the 1970s. Through most of the 1980s, the question of Sami rights to land and water has been the subject of a Government Commission of Enquiry.

In Norway, the Sami Rights Commission submitted its first proposal in 1984. This was concerned with the national legal principles the Commission felt were applicable as the State formalised its relationship to the Sami people. The Commission's proposal contained (among other items):

- the establishment of an independent Sami representative assembly,
- constitutional recognition of the Sami as the aboriginal population,
- introduction of an independent Sami Judicial code.

By the end of the 1980s, despite a change of Government, the Commission's proposal was submitted to and ratified by the Norwegian Parliament.[3] It seems clear that the entire spectrum of the Norwegian political establishment, irrespective of party, acknowledges the need for these policy changes. (Cf. Brantenberg 1991)

In Sweden the question of Sami Rights was the subject of a Government Commission of Enquiry from 1983 until 1990. Towards the end of 1989, the Swedish Sami Rights Commission released its main report.[4] In spite of certain differences of principle between the Norwegian and Swedish

Commissions, the proposal submitted by the Swedish Commission closely resembled the Norwegian proposal submitted 5 years earlier. (Torp 1992) According to the Swedish Commission there were several strategic reasons for this:

"We believe that there is a distinct possibility of creating Scandinavian unity in central questions of policy regarding the Sami. Therefore all, the prerequisites for a common Scandinavian initiative should exist. Few areas of policy would appear so well suited for Scandinavian cooperation as those relating to the Sami."[5]

Actually the Swedish Commission regarded this as a basic precondition for their effort:

"An important precondition for our considerations has therefore been to find a solution appropriate to a unified Scandinavian approach."[6]

Since the Norwegian Commission's proposal had by this time been approved by the Norwegian Parliament with hardly any political opposition, there was a growing optimism among the Sami in Sweden. Now it looked like there was hope of a "new Deal" in Government Sami policy in Sweden.

The Social Democratic Government at that time chose, however, in spite of massive Sami protests, not to submit a collected proposal, which was a necessary condition for a new approach to Sami rights, but submitted instead two limited proposals about forestry in mountain-area forests and changes in reindeer-herding laws.[7] The proposal related to reindeer-herding laws was rejected by Parliament, the right-wing opposition parties being extremely critical of the Government's refusal to submit a collected Bill. When the right-wing parties formed a coalition Government one year later, one of the promises of the new Prime Minister was to submit a collected Sami Bill. Naturally, this aroused hope among the Sami:

"We were waiting and hoping... We were shouting for a collected proposition... We trusted the new Government, and sure enough, there was a collected Sami Bill; we received it with a cheer: Finally!! We leafed excitedly through the massive volume, looking eagerly for the sections on Sami Rights, Language Rights, Constitutional Protection, how the Judicial and cultural situation of the Forest Sami was to be improved, how the economic base of Sami villages was to be expanded, how the demands of the non-reindeer herding Sami for their ancestral lands had been met... But no! We found nothing about any of these."[8]

The most noticeable aspect of the Bill is that the Government does not seem to feel it is necessary to make a clear break or shift in Swedish Sami policy. Rather, it appears that the Government is moving towards a renewal of confidence in the principles the authorities have traditionally followed in their administration of Sami affairs. In other words, the Bill gives the impression that things are to remain precisely as they have always been. In reacting to this, a number of Sami representatives have said that the entire enquiry procedure now seems meaningless and a waste of time, since such a large part of the results of the Sami Commission was missing from the Bill.[9]

And it is reasonable that much of the Sami reaction is related to precisely that point. That the Swedish State had not responded to or acknowledged Sami demands has been a disappointment to many. Indeed, the Government's little felt need to recognize Sami Rights - or accept the long-felt Sami need for national reform - has been seen, by some, as an insult. Others have linked this with Per Unckel's statement that being mentioned in the Constitution is largely irrelevant!

One point in particular in the new Bill caused Sami anger. That is the elimination of the exclusive Sami rights to hunt and fish in "Lapland and those areas of grazing land comprising the reindeer grazing forests".[10] On that point the Bill - the aim of which is to strengthen Sami rights - will in fact weaken them. And it is this proposal that has caused the strongest reactions.

Event 2

The board of the Sami Unity National Congress expressed their complete disappointment with the Government's proposal at a press conference held immediately after the conference opening speech. Lars Anders Bär, the vice-chairman of SSR, said that the Government's suggestions about fishing and hunting were the result of "Agricultural Imperialism" on the part of LRF (the Swedish farmers' organisation)[11] and that the Government's policy would reinstate the paternalistic laws from the 1920s. And he continued:

"A consequence of the Government's proposals would be a nationalization of the traditionally exclusive Sami rights concerning fishing and hunting. This could be compared with Stalin's confiscation of the kulaki's properties in Russia in the 1930s".[12]

It is notable that social scientists have, in fact, already compared Sweden indirectly with the former USSR, since these nations, unlike Norway and Finland, have neither a Sami Parliament nor do they acknowledge the Sami in law. It would seem that both SSR and the Swedish Government have taken note of this comparison.

What then was the reaction to this outspoken protest? Firstly, I shall discuss the Government's reaction, and then the "public" reaction, as it can be interpreted from media commentary.

The Government, through the Minister for Education and Sami Affairs, Mr. Per Unckel, responded directly to these events by releasing the following information to the press for "immediate publication:"

"Statements about the Government and Government policy made by representatives of the Sami population during the last few days seriously disturb existing relations. Organisations whose appointed representatives call the Government policy "Stalinist" and burn Government Bills discredit themselves," says Minister of Sami Affairs Per Unckel commenting on events at the Sami Extra National Congress during the week.

"During the past year my relations to the Sami population have been constructive and stimulating. It is obvious that matters involving different interests could not be brought to a solution that would completely satisfy all parties.

I request most strongly a complete explanation from the Sami representatives, explaining the purpose of the attitude they have chosen to take. An apology for the damage already done would be appropriate".[13]

The Government's reaction, then, was immediate and notably dramatic in tone, for a political context. The day after issuing its press release, the Government received the apology it had requested from the Sami. In a personal letter to Education Minister Per Unckel, SSR Chair Ingwar Åhren "sincerely regrets" the comparison that was made with Stalin's policies. The SSR Chair could, however, find no grounds to apologise for the burning of the Bill, since as he explained, the SSR could not be expected to censor its membership.[14]

Event 3

It would appear that order had been restored, but this was not the case. On the same day that Unckel received his apology, there was a further Sami reaction. As the SSR concluded its congress, another Sami national organization, Same Ätnam, convened its extra National Congress. Same Ätnam, which has established itself as the main organization of the non-reindeer herding Sami, is not - unlike SSR - engaged in direct negotiations with the Government. Here too the Government Bill was the most important item on the agenda. In her opening speech the Chair, Sonja L. Popa, continued Mr. Bär's comparison of Swedish policy in an international perspective. She said:

"In the history of Sweden's Sami policy there has never been anything that could be described as honourable or praiseworthy for a leading democracy. Nevertheless, Sweden would like to play the role of the world's conscience in international conflicts between indigenous peoples and colonial powers.

But now, the Swedish flag is no longer clean; it has stains on it, we better wash it!"[15]

She then produced a bucket full of water and, to the delight of the audience, began washing the Swedish flag.

Her action was highly symbolic, but was still amusing; a good old-fashioned scrubbing of the Swedish flag.[16]

The media response to this action was quite obvious although it was not as vocal as one might expect, considering the Government's rather dramatic reaction. In addition to the various Sami media the events were well covered by local papers, radio and TV. This deliberate use of photogenic political symbols gave a considerable "pay-off" as the burning of the Sami Bill was shown 20 times on the national TV-channel during a period of two weeks.[17] (Cf. Jhappan 1990)

In the Government's press release, the burning of the Sami Bill and the comparison with Stalin are mentioned as being particularly insulting. Neither then nor at any time since has the Government found any reason to comment publicly on the other demands of the Sami - for example the resolutions passed at the National Congresses of the two organizations. To understand the reactions of the Government it is necessary to see them in the context of the general political culture in Sweden.

Swedish political culture

In a historical perspective the political culture in Sweden has solid traditions of being consensus oriented. (Gustafsson 1989) This has influenced the political debate as well as the decisions taken at different political levels.

As a result of the recently completed enquiry on power relations in the Swedish society much research has been produced within the field of political science in Sweden.[18] One of the important findings is that the Swedish political structure is characterised, to a great extent, by an association-oriented democratic ideal. (Olsen 1990) This ideal has as a precondition a well-developed associational structure and a large number of associations. It is unlikely that any other Western democracy has such a comprehensive network of active public associations as in Sweden. (Togeby 1989) In order for this system to function, it is of course essential that citizens express their wishes through these organizations, and not through any ad-hoc movement.

"A precondition for an association-oriented perspective is that people's comprehension of what is good, right and beautiful reflects their common culture tradition. ...This applies to a common understanding of actual conditions, as well as of moral principles and values. ...From an association-oriented perspective, it is essential that preferences and needs, interests and demands are tested and modified in rational discourse in the light of common experience. They must be publicly ratified, depending on the degree to which they are in accordance with what is seen as being best for the group. (Olsen 1990 p 48)

This is a condition that even the Sami organizations have been compelled to live with. Organizations (especially small organizations) which have chosen confrontation as a strategy have found it difficult to have their demands met. Action groups which use hunger strikes and various forms of civil disobedience to gain a hearing have generally not won public acceptance, and are therefore an unusual phenomenon in Sweden. Some people have in fact argued that civil disobedience should be increased in Sweden because Swedish people lack an awareness of injustice and oppression within their society. (Ofstad 1990) In one of the reports of the Enquiry into Power Relationships it says:

"There is no reason to underestimate the degree of unity in "the Swedish democratic ideal". A historic-cultural development has in important ways arrived at common attitudes and opinions that can be appealed to and reasoned with." (Olsen 1990 p 94)

During practically the entire period of enquiry of the Sami Rights Commission, the Sami national organizations operated according to these customary patterns and within the Swedish political system. The only clear exception was in 1983 when SSR, during the initial phase of the Sami Inquiry, refused to co-operate with the Commission, due to the limitations of its directives. Co-operation was, however, resumed within a few months. Since then, contacts between the SSR and public authorities can be described in terms of a willingness to negotiate and a desire to arrive at common solutions. The SSR leadership has been able to construct this strategy with little opposition from their members. In Hirschman's terms, the SSR as an organization has largely been characterized by "loyalty" and hardly at all by "voice" or "exit". (Hirschman 1970).

But the events referred to above have completely broken this pattern. Communication takes place in an entirely different way: it might be said that they have quite simply changed "language".

The first thing one notices is their symbolic significance. Burning the Sami Bill has a quite obvious symbolic meaning. To set fire to something is, in many cultures, the most forceful way of showing disgust and hate. In particular the burning of a flag has high symbolic value and often generates a highly antagonistic response.

The washing of the flag is a more humorous way of showing disrespect and indignation, is more subtle in its message and deflects the antagonism a burning of the flag would have caused.

The statement made by the SSR vice-chair comparing the Government's proposal with Stalinist policies in the 30s also has a very large negative weighting. Throughout the West, the policies of Stalin's Government are a symbol of merciless cruelty and the oppression of human rights. The Swedish Conservative Party has systematically utilized the previous Soviet Communism regime in their party-political rhetoric as the prime example of why Communism should be exterminated. Obviously, comparing Conservative Party policies to Stalinism was rather sensitive!

But all these events have also another level of meaning beyond the more obvious symbolic aspects. The process of signification where actors produce signs and symbols provides information about how actors understand their surroundings. The use of symbols also gives one single actor the opportunity of changing or redefining the definition of the situation. The most powerful way of doing this is through the process of metaphorization and metonymization. (Cf. Thuen 1982)

"The effect is metaphoric when you move a sign to a context where it was not used before, and where it is put to stand - metonymically - for a new whole, a new sequence of events." (Grönhaug 1975:24)

And, as it is argued, this is nothing less than how we attain consciousness:

"Consciousness is seen as the actional operation by which the actor selects elements within fields of perception and acting to let them "stand for" or signify other elements metonymically and metaphorically."
(Grönhaug 1975:27)

Let us now return to the events referred to above. In all the events we find a quite similar process of metaphorization and metonymization. When the vicechair of SSR Mr. Bär compares one specific part of the Government Bill with the politics of Stalin, the metaphoric effect ensures the whole Bill, or the Government's total treatment of the Sami population, to be understood in terms of oppression and annihilation. How the Government, and the media, understood his comparison is de facto evidence of the metonymic success of his comparison.

In the case of the burning of the Sami Bill the key to the metaphoric effect lay in the introductory words; "this is a good Bill if you want evidence of how Sweden treats the Sami". Which means that it is not only the Government Bill that is being burnt, it is also the treatment of the Sami over a long period of time by the Swedish Nation State.

Finally when the chair of Same Ätnam, Mrs. Popa, washes the Swedish flag it is not only a criticism of Government proposals, the metaphoric effect turns it into a criticism of Swedish society as a whole, or actually more fundamentally, everything Swedish! One might say that the effect of these actions together sought to alter the paradigmatic contours where the Sami people form an aboriginal minority within the Swedish Nation State. (Cf. Thuen 1992)

I have not yet answered the question why the Government reacted so strongly to these events. For this purpose I will attempt to analyze these events, emphasizing the communicative aspect.

A generally accepted formulation in social science is that "the message consists of more than the words that are spoken". Many of the messages we exchange tell us how other messages are to be interpreted. This is what Bateson calls metacommunication: messages about messages. With metacommunication we put "frames" around messages that tell others how to interpret them. Bateson says:

"We start, then, with a potential differentiation between action in context, and action or behaviour which defines context or makes context intelligible. A function, an effect of the metamessage is in fact to classify the message that occurs within its context." (Bateson 1980 p 129)

All of the actions referred to above have a common metacommunicative aspect, or, in Bateson's terms, they define a common context. In all cases, the metamessage is quite clear. It says, "We, the Sami people, will not respect the traditional negotiating procedure that, until now, we have been forced to accept. In the future we shall feel free to present our demands through the channels we find most receptive!" Which, as we have seen, on these occasions was through the media.

The rapid response from the Government shows quite clearly that they comprehended the metacontent of the message. Their blunt and threatening reaction indicates that the Government felt the message to be a serious

137

threat. It is reasonable to assume that the Government actually saw the new Sami ethnopolitical strategy as of mortal danger. Increased media attention would probably result in the following:

a) A broad public debate of the Government Bill which would reveal that the Government Bill falls far short of the Sami Commission's proposals.

b) That Sweden trails far behind other Scandinavian countries in confirming that the Sami are the aboriginal population.

c) Opinion might turn against Government policy in these questions - just as happened in Norway with the Alta/Kautokeino hydroelectric project.

d) That the debate on the Swedish Sami policies would also attract the attention of international media and seriously damage Sweden's reputation abroad - in the UN, for example.

Above all, the development of negative international opinion could have serious repercussions for the Swedish Government. There are several examples of aboriginal groups changing the political situation in their own countries through successful international campaigns. (Cf. Paine 1985 and Ponting 1990)

Thus it was of decisive importance for the Government to stop any move towards a new Sami strategy of negotiation, or different form of communication. By forcefully threatening a possible break in negotiations, the Minister of Sami Affairs, Per Unckel, successfully averted such a development. The Government received the apology it had demanded, and thereby "a guarantee" from the SSR that nothing similar would happen in the future. The Sami, however, are caught in "a Double Bind" where they cannot win whatever they do. They have not attained satisfactory results through "polite" negotiations and they will be cut off the agenda if they engage in any protest action. (Bateson 1978)

Concluding remarks

In conclusion, we can say that the Sami protest actions referred to here were of a dramatic symbolic nature and consequently of major media potential. This was despite their being performed in what we can rather loosely call Sami contexts. At the same time, we can confirm that these actions resulted from the spontaneous and indignant reactions of individuals to Government proposals and were not part of a conscious political strategy on the part of the Sami organizations.

There are several arguments in support of this position. In the first place the time was hardly ripe for the Sami to introduce a new negotiating strategy. After ten years of peaceful negotiations only a few weeks remained before the Government's proposals were to be dealt with by Parliament. Even more indicative is the apologetic and hopeful response of the Sami organizations to the Government's threat to withdraw from the negotiations. Had the Sami wished to break off negotiations, with the object of gaining support for their ethnopolitical demands, then the situation engendered by the Government's unexpectedly strong reaction, provided them with a "golden opportunity." If the Government had persisted in its demands and broken off negotiations, this would, in itself, have given the Sami immediate public opinion advantages and aroused the interest of the international mass-media. That the largest Sami organization should be "deeply apologetic" in the situation that had arisen can only be interpreted as indicating that the actions considered in this article were not part of a strategic plan on the part of the Sami political organizations. In retrospect we can, however, ask ourselves, how much the guarantees made by the SSR are in fact worth!

The day after the Government had issued its first warning via a press release, Sonja L. Popa scrubbed the Swedish flag at Same Ätnam's Extra National Congress. Thus the Government's threat of "disturbed relations" had very little effect on Sonja J. Popa who did not see herself as bound by the apology the SSR chair had sent to the Minister of Sami Affairs. The fact that many Sami supported Olof Johansson's actions and at the same time that a number have been critical of being so respectful to "the Swedes" - by that demeaning apology - indicate that many Sami would advocate a more

139

confrontational ethnopolitical strategy in the future. The reason for this is simply that many among the Sami people feel they have little, or nothing left to loose. In part this was because the Government Bill was a mere shadow of what the Sami Commission had suggested, and partly because the Sami in Sweden see themselves as an increasing disadvantage in comparison with their "brothers and sisters" in Norway and Finland.

At the end of December 1992 the Sami Bill was passed, with no important discussion, by the Swedish Parliament. The Swedish Nation State has thereby decided, "once and for all", how to deal with the "Sami Problem".

Notes

1.   Swedish Television, Channel 1, "Aktuellt" 18.00 h. 1992.10.22.
2.   Swedish Television, Channel 1, "Aktuellt" 18.00 h. 1992.10.22.
3.   Changes in the Norwegian constitution required a 2/3 majority of two separate Parliaments.
4.   All together the Commission produced three reports:
     a) Statens Offentliga Utredningar 1986:36 Samernas folkrättsliga ställning,
     b) Statens Offentliga Utredningar 1989:41 Samerätt och sameting,
     c) Statens Offentliga Utredningar 1990:91 Samerätt och samisk språk.
5.   Statens Offentliga Utredningar 1989:41 Samerätt och sameting, p 116.
6.   Statens Offentliga Utredningar 1989:41 Samerätt och sameting, p 115.
7.   Regeringens proposition 1992/93:32 Samerna och samisk kultur m.m.
8.   Samefolket (monthly Sami newspaper) no 11, 1992.
9.   Cf. Samefolket no. 11, 1992.
10. Reindeer herding law 1971:437 para 25. (Cf. Prop. 1992/93:32 Samerna och samisk kultur m.m., p 131)
11. The lobbying of the farmer organisations is confirmed by the former Minister of Justice Laila Freivalds, see Torp 1991.
12. Samefolket no 11, 1992.
13. Press release, Education Ministry, 92.10.23
14. Samefolket no 11, 1992.
15. Samefolket no 11, 1992.
16. The event was later reported to the police as being defamatory to the Swedish flag. The charge was dropped, however, for Swedish law does not forbid washing the flag!
17. Samefolket no 11, 1992.
18. The Commision of Enquiry was working from 1985 untill 1990.

References

Bateson, G., 1978, *Steps to an Ecology of Mind*, London.

Bateson, G., 1980, *Mind and Nature - A necessary unit*, Glasgow.

Brantenberg, T., 1991, Norway, *Constructing Indigenous Self-Government in a Nation State*, in Jull et Roberts (eds.): The Challenge of Nothern Regions, Australian National University, Darwin.

Grönhaug, R., 1976, *Transactions and Signification: An Analytical Destinction in the Study of Social Interaction*, Mimeo, Dept. of Social Anthropology, Univ. of Bergen.

Gustafsson, H., 1989, *Sockenstugans politiska kultur*, Stadshistoriska Institutet, Stockholm.

Hirschman, A.O., 1970, *Exit, Voice and Loyalty*, Cambridge.

Jhappan, C.R., 1990, Indian Symbolic Politics: The Double-Edged Sword of Publicity, *Canadian Ethnic Studies/Etudes Ethniques au Canada*, vol. 22 no. 3. Canadian Ethnic Studies Association, Saskatoon.

Ofstad, H., 1990, Svenska folket är för lydigt, Debattartikel *Dagens Nyheter* 1990.11.21.

Olsen, J.P., 1990, *Demokrati på svenska*, Stockholm.

Paine, R., 1985, Etnodrama and the "Fourth World": The Saami Action Group in Norway, 1979 - 81, in Dyck, N., Indigenous Peoples and the Nation state: Fourth World Politics in Canada, Australia and Norway, *Social and Economic Papers* 14, Inst. of Social and Economic Research, Memorial Univ. of Newfoundland, St John's.

Ponting, J.R., 1990, Intenationalization: Perspectives on an Emerging Direction in Aboriginal Affairs, *Canadian Ethnic Studies/Etudes Ethniques au Canada*, vol. 22 no. 3. Canadian Ethnic Studies Association, Saskatoon.

Thuen, T., 1982, *Meaning and Transaction in Sami Ethnopolitics*,ISV, Occ. Papers, University of Tromsö, Tromsö.

Thuen, T., 1982, *Quest for Equality - Norway and the Sami Challenge*, ISV, Occ. Papers, University of Tromsö, Tromsö.

Togeby, L., 1989, *Ens og forskellig - gresrodsdeltagelse i Norden*, Politica.

Torp, E., 1991, Svenska regeringens behandling av samerättsutredningen, Retfärd; *Nordisk Juridisk Tidsskrift* no 53, vol 14, Copenhagen.

Torp, E., 1992, Sami Rights in a Political and Social Context, in, Lyck, L. (ed.): *Nordic Artic Research on Contemporary Artic Problems*, Copenhagen.

Collective Entrepreneurship and Microeconomies

Ivar Jonsson

The main topic of this paper is to analyse the fundamental dynamics of entrepreneurial activity in relation to the particular size related problems of accumulation of microeconomies. In the first part entrepreneurial activity is defined and neo-classical theory is criticized for it's lack of realist account of entrepreneurial activity. An alternative theory is offered which emphasises the collective nature of entrepreneurship and is based on an evolutionary and institutionalist theory of technical change. The concept of collective entrepreneurship is defined which takes into account social as well as economic determination of technical change. In the second part, the constraints of entrepreneurial activity are analysed in terms of the particular size related problems of accumulation of microeconomies. In the third part, entrepreneurial activity is analysed in terms of increasing globalization trade and production and global firms. Finally, in the last part problems of collective entrepreneurship in Iceland as a microeconomy are observed.

Economic growth, search strategies and techno-economic paradigms

Entrepreneurial activity is essential for economic growth. Research into the causes of economic growth have shown that, unlike what orthodox, neo-classical theory presumes, it is competitiveness in research and development (R&D) and capacity to deliver rather than competitiveness in labour costs per unit, that is important for economic growth in the medium and long run (see J. Fagerberg 1988 for a study of economic growth of 15 OECD countries during the period 1961-83). The dynamics of technical change should therefore be of great interest for students of economic growth. However, this is not the case. Research into the dynamics of technical change requires institutional approach that does not only take into account economic conditions of technical change but highlights as well the role of political structures and actors and sociological aspects dealt with in organizational theory. Studies of economic growth have been predominated by neo-classical theory. This train of thought suffers from serious shortcomings as concernes technical change. Its inability to deal with institutional determination of technical change is its greatest limitation in this field of study.

In neo-classical theory, firms are viewed as operating according to a set of decision rules that determine what they do as a function of external (market) and internal (such as available capital stock) conditions. These rules are reduced to the principle of *maximzing* on the part of the firms, which usually refers to maximizing profit or present value of the firms. To be able to calculate maximum output in this sense, firms are further presumed to have precise knowledge of how to do. In terms of production in the traditional sense, such a precise knowledge of how to do refers to maximizing activities or techniques and consequent "production sets". In terms of other fields, such a precise knowledge of how to do refers eg. to advertising policies or financial asset portfolios. Finally, it is presumed that maximizing firms make their decisions or choices on the basis of given sets of known alternatives to choose from, whether these are alternative actions, market constraints, internal constraints such as short term available quantities of factors. In some models, the idea of maximizing behaviour takes into account information imperfections, costs, and constraints (R.R. Nelson and S.G. Winter 1982, 12).

The neo-classical principle of maximization is inadequate as a microlevel explanation of how firms make their decisions and choises as concerns technical change. The real world is much more complicated and much more uncertain than neo-classical theory would have it. In periods of technical change, maximization in the literal sense becomes very difficult to say the least, because knowledge of how to do is undermined as competitors exploit new technology which is not yet diffused and unknown to the firm in question. As a consequence, the neo-classical postulate of choosing between known alternatives and maximizing becomes unrealistic. Furthermore, industrial R&D, invention and innovation is by nature open ended and results are to a high degree uncertain. A more sofisticated concept is needed to analyse firms decision rules than the concept of maximization.

The concept of 'routine' supersedes the naive formalism of neo-classical theory (R.R. Nelson and S.G. Winter 1982, 14-18). This is a concept developed by economists and is is related to the sociological concept of social norms, but is comparatively underdeveloped. The concept of a 'routine' refers to all regular and predictable behavioral patterns of firms and covers characteristics of firms that range from well-specified technical routines for producing things, through procedures for hiring and firing, ordering new inventory, or stepping up production of items in high demand, to policies regarding investment, R&D, advertising, and business strategies about product diversification and overseas investment (ibid., 14). R.R. Nelson and S.G. Winter have identified three main types of routines depending on their different levels of abstraction of decisions making. Firstly, there are routines that refer to what a firm does at a any time, given its prevailing stock of plant, equipment, and other factors of production. These are routines that govern short-run behaviour and have been called 'operating characteristics'. Secondly, there is a set of routines that determine the period-by-period augmentation or dimunition of the firm's capital stock, as an example when it is decided whether to implant a new machine or repair an old one, building new plant or investing in a major R&D program on a recently opened technological frontier. Thirdly, there are routines that operate to modify over time various aspects of the operating characteristics of firms. These are routines that guide the 'searches'' of firms as they change the routines mentioned above. Search policies or strategies of firms are determined by routines that take into account different factors, such as size of the firm, anticipated level of risk and profit, what competitors are doing, assessment of the payoff of R&D in general and of classes of projects in particular, evaluation of the case or difficulty of achieving certain kinds of technological advances, and the particular complex of skills and experience that the firm possesses (ibid., 16-18 and 249).

Studies on the decision making process of R&D in firms reveal how unrealistic the neo-classsical theory is with its maximizing principle. The decsion making process reflects the uncertain nature of R&D and technical change. It appears from the studies that a widely used procedure is to begin by developing lists of projects that if successful would have high payoff, and then screening this list to find those projects that look not only profitable if they can be done, but doable at reasonble cost. Payoff-side factors are examined first, and those relating to cost or feasibility are looked at second. However, in certain search, R&D and industrial innovationfirms proceed by focusing first on exciting technological possibilities and then screening these to identify the ones that might have high payoff if achieved. Neither case is literally optimal. Since all alternatives cannot be considered, there must be some rather mechanical procedures employed for quickly narrowing the focus to a small set of alternatives and then homing in on promising elements within that set (ibid., 255).

Furthermore, search, R&D and industrial innovation is not simply a matter of responding to market demand. The role of the selection environment has to be taken into account and it can not simply be reduced to market-demand. The market determines search, R&D and industrial innovation in so far as competition forces firms to imitate and exploit new technology that reduces production costs and hence prices. Those firms that do not follow this rule perish from the market. However, the relations between markets and firms are not alltogether one-sided. Typical market structures are not perfectly competitive and firms try to modify the demand of their products by employing advertising and research and development as a central commpetitive weapons (Packard 1975 and Galbraith 1967).

There are other nonmarket selection environments as well. Most theorizing of market selection presumes a relatively clear separation of the "firms" on the one hand, and consumers and regulators on the other. ˙Consumer evaluation of products - versus their evaluation of other products and versus price - is presumed to be the criterion that ought to dictate resource allocation.   Nonmarket sectors are not characterized by such clear seperation between firms interests and consumers interests. Search, R&D

and industrial innovation is affected by more complicated set of criteria than maximazation of firms' profits and consumers' utility in market terms. In the case of a public agency such as a school system, and its clientele (students and parents) and sources of finance (mayor, council, and voters) there is not the arm's-length-distance as between a seller and buyer of a car. The public agency is expected to play a key role in the articulation of values and to internalize these and work in the public interest. Even in nominally private-sector activity such as in the provision of medical services, doctors are not supposed to make decisons regarding the use of a new drug on the basis of the profits he or she makes from it. To mention but few example of nonmarket selection environments, we would highlight public regulation concerning pollution and public health standards, the public postal services and ministries of defence that affect search, R&D and industrial innovation through procurement etc (R.R. Nelson and S.G. Winter 1982, 268-72).

Sofar, we have analysed the conditions of economic growth by observing the micro-level principles of firms activities as concerns technical change. However, the routines of firms and search strategies develop in macro socio-economic contexts that generate and reproduce clusters of basic ideas or paradigms that mould the micro-level routines and search strategies of firms. Such clustures of basic ideas have been called 'techno-economic paradigms. Techno-economic paradigms refer to clusters of ideas in the field of organization of production and technical change that change the basis range of industries and generate technological revolutions and long lasting economic waves, i.e. Kondratieffs. Fordist mass production was the key technological factor that generated the long wave of economic growth between the 1930/40s to 1980s, while information and communication technology is the key technological factor in the present process of shifting techno-economic paradigms.

Diffusion of techno-economic paradigms depends on changes in regimes of accumulation, i.e. social and political structures that foster, reproduce and transform basic ideas of best practice technology and organization of work. Such changes of regimes of accumulation depend on the balance of power of social and political forces and their struggle in the process of hegemonic politics (see Jonsson 1991a). In this sense changes of techno-economic paradigms differ from other classes of innovations as they are inter-regime changes. From these changes we can distiguish innovations that lead to intra-firm changes, inter-firm changes and inter-branch changes. Intra-firm changes refer to innovations that change the technological base of individual firms and their organization in an incremetal way. They often occur, not so much as the result of any deliberate research and development activity, but as the outcome of inventions and improvements suggested by engineers and users of technology ('learning by doing and learning by using). Inter-firm changes refer to changes in the relations beween firms in the sense that new products are produced. They are based on adical innovations that usually are the result of deliberate research and development activty by enterprises and/or university and government laboratories. Unlike incremental innovations, they not occcur from improvement of existing processes or products of production. Rayon as an example could not have resulted from the improvement of rayon plants or the woollen industry. They are important as the potential springboard for the growth of new markets. Radical innovations are relatively small and localized but may develop over a period of decades into new industries if clusters of radical innovations are linked together as in the case of the synthetic materials industires or the semiconductor industry. Finally, Inter-branch changes refer to changes of technological systems that are far-reaching changes in technology, affecting several branches of the economy, as well as giving rise to entirely new sectors. They are based on combination of incremental and radical innovations, together with *organizational* and *managerial* innovations affecting more than one or a few firms. An obvious example is the cluster of synthetic materials innovations and petro-chemical innovations (Freeman and Perez 1988).

All types of innovations and all classes of technical change are generated through the process of entrepreneurial activity. This process is complicated and can not be reduced to the avctivity of an individual entrepreneur.. Let's look closer at the matter.

Entrepreneurship

The concept of the entrepreneur has had a "come back" in economic discourse in recent years after having been trivialized by neo-classical theory. Neo-classical theory trivialized the concept with its emphasis on perfect information and perfect markets according to which the entrepreneur plays a static and passive role which was reduced to the efficient size of the firm and marginal efficiency curves. Such a view has some relevance in periods when economic development is relatively stable and profitability and productivity forecasts etc. can be based on past market trends.

However, neo-classical theory fails both in its emphasis on marginal efficiency as a guiding principle in running business and in its a-historical approach as economic and social uncertainty affects the rationality of investment and such uncertainty is periodic due to long waves, i.e. Kondratieffs, and technical and social change (Jonsson 1991b). When a long wave in the world economy enters the phase of a recession and markets become saturated, profits have been competed away and a shift to a new techno-economic paradigm is necessary (Freeman 1987), the role of the entrepreneur becomes the more important. It goes for all periods that the role of the entrepreneur is to make judgemental decisions, .i.e. to take managerial decisions when no decision rule can be applied that is both obviously correct and involves only freely availble information (Casson 1982). However, the uncertainty level of judgemental decisions is historically determined as uncertainty is greatest at the lower and upper turning point of long waves. But, there more critical points to be made concerning this definition of entrepreneurial decisions.

There are two critical points that should be emphasized concerning the definition above of an entrepreneur. Firstly, all management decisions presume judgemental content and uncertainty. As a consequence, one has to distinguish between basic entrepreneurial decisions and other management decisions. An entrepreneurial decision is different from other decisions insofar as it is related to the realization of the entrepreneurial function. We will discuss that function below. Secondly, the idea of an entrepreneur presumes that it is an individual or a firm that makes entrepreneurial decisions. This view is a myth as, on the one side entrepreneurial activity is as much a product of the accumulated knowledge and technological progress of the society that hosts the entrepreneur as a product of his/her insight. On the other side, it is a myth as firms are not totally unified entities and decisions by firms are a product of conflict ridden processes in which different departments and individuals on different managerial levels take part. Furthermore, managerial decisions in large modern firms are taken by teams rather than individuals and as such they are more than sums of the opinions or ideas of the individuals in question.

The idea of the entrepreneur as an isolated genius is misleading and innovations do not fall on the heads of individual entrepreneurs as manna from heaven. In fact, entrepreneurship is a social process in which innovations are generated by social and cultural conditions that constitute at the same time the preconditions of their establishment and acceptance (G. M. Hodgson 19, 268). Realist analysis of entrepreneurial activity require a qualitative research into the structural conditions of techno-economic as well as social innovations. Following J. A. Schumpeter, we would claim that entrepreneurial activity centers around realizing the entrepreneurial function. As Schumpeter puts it:

"...the function of entrepreneurs is to reform or revolutionize the pattern of production by exploiting an invention or, more generally, an untried technological possibility for producing a new commodity or producing an old one in a new way, by opening up a new source of supply of materials or a new outlet for production by reorganizing and industry and so on" (Schumpeter, 132).

Schumpeter's concept of the entrepreneurial function is inadequate as it does not take into account social innovations. His concept reduces innovations to pure economic and technological factors, while social innovations are bypassed. By social innovations we refer factors such as developing new consumer tastes and traditions, transforming the knowledge base of nations, restructuring industrial relations, organizing new systems of interest mediation, generating firm-nets, user-producer relations, new forms of

152

interlocking directorships, etc.

Fundamentally, social innovations affect external economies of scale, while technological or Schumpeterian innovations affect internal economies of scale. Briefly, economies of scale can be analysed as internal plant and internal firm economies of scale and furthermore as external plant and external firm economies of scale. Internal economies at the plant level derive from the exploitation of production techniques involving the specialization of labour, machinery and management and the accumulation of knowledge through experience in the production or running the plant. Internal firm economies on the other hand refer to the scale of management, distribution, the acquisition of inputs and the organization of research and development facilities. External economies at the plant level refer to factors such as access to credit and to cheap inputs of wage goods and capital equipment resulting from easy access to other suppliers and from available social infrastructure (communication, education, research, law and order, etc.). External firm economies, finally, refer to the access to large scale credit facilities, and to the communications and social and educational services required to maintain a high level of manpower division and be able to sustain control of international dimensions such as access to foreign markets and capital. These economies are sectorally, spatially and historically unevenly distributed so that locational specificity measured in terms of the distribution of internal and external economies, determines the different volumes and rates of capital accumulation which generates uneven economic development between countries, monopolization and disequilibrium (Brett 1983 and Jonsson 1991a).

It is clearer today than ever before that the entrepreneurial agent that realizes the entrepreneurial function can not be reduced to an individual or firm as we mentioned above. The active economic role of institutional actors such as local authorities, communes, the central state, international organizations and organized cooperation between firms in regulating and promoting conditions of competition and competitiveness leads us to the conclusion that the entrepreneurial function is realized through a process of collective entrepreneurship. This is a process in which external economies

153

of scale are created and transformed. Furthermore, as the institutional base
of economic activity is different in the different countries, so is the
organization of collective entrepreneurship different. However, collective
entrepreneurship is not limited to national economies alone. Increased
foreign direct investment (FDI) and globalization of capital accumulation
(Chesnai 1988 and Julius 1990) due to cross border activities of
multinational corporations (MNC) has generated forms of collective
entrepreneurship that are essential for competitiveness of firms in
international markets such as automobiles and electronics (K. Hoffman and
R. Kaplinsky 1988 and Chesnai 1988).

Forms of collective entrepreneurship

Basically, forms of collective entrepreneurship depend on the one side on
the entrepreneurial actors and, on the other side, on structural constraints,
i.e. 1) constraints that are country specific such as the constraints of a
microeconomy and 2) constraints set by the structural development of
international trade and the world market.

As concerns entrepreneurial actors, forms of collective entrepreneurship are
based typically on relations between actors such as the state, organized
interests, firms and individuals. The following table 1 highlights some well
known forms of relations between these actors that can be considered as
examples of collective entrepreneurship. Table 1 indicates that there are
many possible forms of collective entrepreneurship. Depending on the
balance of power between the entrepreneurial actors and depending on
culture and history of different countries, the hegemonic role in
entrepreneurial activity may be played by the state such as in the case of
'state entrepreneurship' in Taiwan and South-Korea (Davis and Ward 1990,
Huang 1989 and Cotton 1992). In other cases, such as in USA, firms and
markets are more important in determining the path of innovative activity
(Nelson 1988).

154

Table 1.

Forms of collective entrepreneurship

Actors :

The state : governments, municipalities and institutes - The State :
- Supporting international R&D projects, e.g. EUREKA, ESPRIT, ERASMUS; developmental plans for R&D on regional level; establishing R&D funds and institutions, science parks etc.; providing tax reductions for R&D, procurement etc.(see table 2 for more details)

The state : governments, municipalities and institutes -
Organized interests of capital and labour :
- Collaboration between employers organizations, trade unions and the state in developing R&D and innovative institutes run by organized interests.

The state : governments, municipalities and institutes - Firm :
- Procurement, science parks, tax allowances for innovative firms etc. (see table 2 for more details)

The state : governments, municipalities and institutes - Individual :
- Centres and laboratories for inventive individuals.

Organized interests of capital and labour - Firm :
- R&D funds and institutes established and run by employers organizations and/or trade unions from different branches of industry.

Organized interests of capital and labour - Individual :
R&D contracts with individuals and access to laboratories and other facilities.

Firm - Firm :
R&D and innovation networks of firms; user-producer relations.

Individual - Individual :
Groups of individuals initiate and finance R&D projects.

In table 2, we have highlighted the main types of measures that fall under state forms of collective entrepreneurship as they have appeared in the advanced capitalist countries in recent decades.

Table 2.

State forms of collective entrepreneurship

Policy tools	Examples
1. Public enterprise	Innovation by publicly owned industries, setting up of new industries, pioneering use of new technique by public corporations, participation in private enterprise
2. Scientific and technical	Research laboratories, support for research associations, learned societies, professional associations, research grants
3. Education	General education, universities, technical education, apprenticeship schemes, continuing and further education, retraining
4. Information	Information networks and centres, libraries, advisory and consultancy services, databases, liaison services
5. Financial	Grants, loans, subsidies, financial sharing arrangements, provision of equipment, buildings or services, loan guarantees, export credits, etc.
6. Taxation	Company, personal, indirect and payroll taxation, allowances
7. Legal and regulatory	Patents, environmental and health regulations, inspectorates, monopoly regulations

8. Political	Planning, regional policies, honours or awards for innovation, encouragement of mergers or joint consortia, public consultation
9. Procurement	Central or local government purchases and contracts, public corporations, R&D contracts, prototype purchases
10. Public services	Purchases, maintenance, supervision and innovation in health service, public building, construction, transport, telecommunications
11. Commercial	Trade agreements, tariffs, currency regulations
12. Overseas agent	Defence sales organizations

Source: R. Rothwell and W. Zegweld, 61.

Microeconomies and collective entrepreneurship

Collective entrepreneurs face different constraints upon their activity depending on country specific conditions of capital accumulation. Microeconomies, i.e. economies with less than one million inhabitants, have particular size related problems of accumulation different from large economies (Jonsson 1991a and 1992). These size related problems of accumulation in microeconomies appear in five fundamental structural constraints of capital accumulation: 1) the absolute number and size of firms tends to be very small in microeconomies (as the case of Iceland indicates, see table 3 and the Appendix); 2) the very small size of the home market; 3) great openness of the economy; 4) great fluctuations in GDP and; 5) the very small absolute size of administration.

Table 3.

Size of the manufacturing sector (ISIC 31-39) and level of value added in the Nordic countries 1985

	Manufacture# (ISIC 3) as % of total labour force	Value added of ISIC 3# as % of gross output persons engaged per establishment	Average size of establishments
Denmark	20.1	40.5	57.9
Finland	21.8	36.8	69.3
Iceland	14.6 (13.8)* (8.0)+(0.8)++	32.9 (34.2)* (16.1)+(23.8)++	8.5
Norway	17.9	27.2	44.0
Sweden	22.4	41.2	84.6

The figures on Iceland are based on The Economic Institute. Other figures are based on Nordic Council of Ministers 1988 and OECD 1989.
* ISIC 31-6 and 38-9, i.e. ISIC 30 and 37 excluded (ISIC 37 = aluminum and ferro-silicon production by multinationals).
+ ISIC 30
++ ISIC 37
Sources: Nordic Council of Ministers 1988; OECD 1989; The Economic Institute.

Furthermore, a microeconomy is characterized by a small home market in absolute terms and the smaller the home market/s is/are, the fewer firms can be established in the markets and firms will tend to be small and threatened with over-investment due to difficulties of exploiting economies of scale. The smaller the economy is, the more unlikely it is to be self-sufficient in terms of production of goods demanded (depending on the diffusion of markets and consumption of industrially produced goods). As a consequence, the smaller the economy is, the more open it must be. The smaller the economy is, the greater will the fluctuations in GDP be. This is the case because the smaller the economy is, the fewer the branches of industry are. Thus, fluctuations in one part of the economy may not be met by counter-affecting fluctuations in other parts of the economy as is the case in larger economies. Finally, the smaller the economy is, the smaller is the administration in absolute terms. The size of the administration constrains its neutrality, quality, forms and way of conduct.

Finally, the very small size of the home market and the small absolute number and size of firms (whether in terms of income or person years) determines monopoly tendencies and chances of exploiting economies of scale. These two last mentioned factors affect levels of value added as monopoly and oligopoly lead to decreased output of the economy and increasing costs of other non-monopoly sectors (Yarrow 1985) and lack of economies of scale leads to relatively low levels of productivity. Furthermore, the openness of the economy affects the role of exchange rates. The smaller the economy is the more open it will tend to be and the more important exchange rate policies will be for the economy. This is the case both in terms of costs of imported goods for consumption and production as well as in terms of profitability of export sectors and long term rationality of investment in these sectors. The fourth factor, fluctuations in GDP, affects social and political stability. Fluctuations in GDP lead to fluctuations in income distribution and class relations as well as fluctuations in state revenues and party voting. Finally, the fifth factor, the absolute size of administration, determines its grounds to function as a formally neutral body vis-á-vis social and economic interests and to contribute to collective policy making. The smaller the administration is in terms of number of persons, the more it is likely to depend on short term influences of governments and interest groups. The smaller the administration is, the more likely it is to lack resources and specialization to contribute to long term policies and economic and political stability (Jonsson 1991a).

The consequence of the size related factors is that the size of an economy affects the resources and level of social, economic and political stability upon which the quality and time-scale of economic policies and accumulation strategies depend (Jonsson 1991a).

The size related problems of accumulation in microeconomies set limits to collective entrepreneurs located in such economies. This is a challenge that collective entrepreneurs need to overcome with special measures. Due to the low level of value added and scant R&D of individual firms and limited possibilities of user-producer networks as firms and branches of industry are few in absolute terms, ineffective use of R&D resources and the risk of investing in industrial innovations is high. Furthermore, as firms tend to be very small in microeconomies, problems of crossing minimum capital thresholds in R&D and lack of marketing new products is severe.

As a consequence, due to all these constraints the need to rationalize entrepreneurship and the need to develop productive systems of collective entrepreneurship is even greater in microeconomies than in large economies. Two principles appear to be necessary cornerstones of strategies to develop such systems in microeconomies: On the one side, it has to be based on country specific know-how in order to develop firms and branches of industry that are able to enter particular market niches; On the other side, in many cases it has to be based on collaboration between domestic and foreign and/or multinational corporations in order to decrease risk and provide the small domestic firms with access to relevant components as well as marketing channels.

Due to the very small homemarkets of microeconomies they tend to be very open in terms of imports and exports compared to large economies (see Appendix). The small homemarkets tend to be to small for new products and they are quick to saturate. The need to export new products is therefore great already in the early stages of the life cycle of products. As a consequence the constraints of the development of international markets is great for entrepreneurial activity in microeconomies.

Globalization and collective entrepreneurship

Increasing international trade is immanent to capitalist development. In the post war era the process of internationalization of capital accumulation has been characterized by higher growth rates of international trade compared to the rate of growth of individual OECD countries (OECD 1992). The first phase of high growth of international trade in that era took place in the 1950s -70s. It followed, on the one side, increased foreign direct investment (FDI) by multinational corporations (MNCs) that increasingly exploited cheap labour in Third world countries. In the 1960s American MNCs invested increasingly in Europe. This increased FDI of MNCs in Europe was followed by increased intra-industry trade between their units or plants in different countries. FDI in Europe decreased again in the 1970s due to the oil crises of the early and late 1970s (Julius 1990). On the other side liberalization of international trade followed agreements such as GATT and international organizations such as EC and EFTA. A second phase started in the 1980s with increased globalization of firms, systemofacture and increasing protectionism in the advanced capitalist countries (Hoffman and Kaplinsky 1988 and Julius 1990). Trade between the advanced capitalist countries has increased fast and FDI in these countries has grown faster than in Third world countries (Julius), but FDI in Eastern Europe and Mainland China increased fast in this period as well (Chesnai 1988).

Global firms characterized by systemofacture - i.e. internationally integrated firms with geographically dispersed units of design, production and marketing, but integrated by the means of information technology - have typically developed in the automobile and electronics industry, but flexible specialization for specified markets is also possible in other industries/markets as the well known cases of Benetton and fashion clothes shows.

Increasing international trade, increased FDI and globalization of firms create both new opportunities and constraints for collective entrepreneurship in microeconomies. This development creates opportunities for easier axcess to market niches and chances increase for small firms in microeconomies to become sub-contractors of global firms. However, the competition between sub-contractors undermines their position vis-á-vis the global firms. Furthermore, small firms have weaker position in competing with big firms as it is more difficult for them to exploit new technology than is the case with big firms. Research into transfer of technology shows as an example that MNCs invest more intensively in information technology then national firms and big firms are more information technology intensive than small firms (Kaplinsky 1984). In this situation the need for productive collective entrepreneurship is felt even more greatly in microeconomies because of the small size of firms.

We may conclude that the present structure of international trade is both an advantage and disadvantage for microeconomies. Liberalization of international trade is necessary for the export-oriented microeconomies and insofar as they can develop and produce high quality products for market niches, liberalization is an advantage for them. However, difficulties in exploiting new technology is a special problem as we mentioned above. The threat of becoming a low tech and low value added sub-contractor of MNCs is great.

Collective entrepreneurship and the case of Iceland

There are three islands in the Atlantic North-Arctic area that can be defined as microeconomies, i.e. the Faroe Islands, Greenland and Iceland. They are all characterized by tiny population and they all are predominantly fish exporting economies with very small and stagnant manufacturing sectors (see Jonsson 1992 for a comparison of these economies and larger economies).

We can not deal with all these three economies in this short paper and will therefore concentrate on the case of Iceland. Being a microeconomy, entrepreneurial activity in Iceland faces structural constraints that are both due to the problem of size for capital accumulation as well due to the constraints of international trade.

In terms of domestic constraints and the problem of size, Iceland is characterized by very small firms and low levels of value added (cf. table 3 above). The regime of accumulation in Iceland has been predominantly extensive for all this century. Economic growth has, on the one side, been based on increasing fishing and fish processing and flow of cheap labour from the country side to the fishing communes and the capital during the first half of the century and from the fishing communes to the capital area during the post war era. Finally increased participation of women in the labour market since the 1970s provided further cheap labour. On the other side economic growth has been based on imported technology for production both in the fishing and fish processing sectors (the F-sector) and in the manufacturing sector. As a consequence, exploiting the technology gap has been a very important factor of economic growth (for detailed analysis, see Jonsson 1991a. See also Jonsson 1992 and Jonsson and Jonsson 1992 for a short discussion).

By importing best practice technology and periodically restructuring the F-sector (importing British trawlers in the late 1940s, stern trawlers in the 1970s and freezing trawlers in the 1990s) and by periodically extending the territorial waters (4 miles in 1952, in 12 miles 1958, 50 miles in 1972 and 200 miles 1975), the extensive regime of accumulation has succeeded in securing long periods of high rates of economic growth. However, in the latter part of the 1970s and during the 1980s growth rates contracted as table 4 shows and the five years period since 1987 has been the longest permanent recession since the war. R&D has never been important for entrepreneurial activity in Iceland as expenditure on R&D has always been less than 1% of GDP (Jonsson 1991). This is to be expected in an economy which is predominantly characterized by an extensive regime of accumulation.

Table 4.

Real Growth of GDP per Capita

	Denmark	Germany	Iceland	UK	USA	OECD
1977	1.3	3.1	7.9	2.3	3.4	2.9
1978	1.2	3.1	6.5	3.7	4.0	3.3
1979	3.3	4.1	4.1	2.7	0.9	2.4
1980	-0.6	0.7	4.8	-2.3	-1.3	0.3
1981	-0.8		2.8	-1.4	1.2	0.9
1982	3.1	-0.9	0.7	1.8	-3.6	-1.0
1983	2.6	1.9	-5.1	3.6	2.9	2.0
1984	4.4	3.2	2.5	2.0	6.2	4.0
1985	4.2	2.1	2.9	3.3	2.8	2.8
1986	3.5	2.2	6.6	3.7	2.2	2.3
1987	0.2	1.4	7.3	4.5	2.5	2.7
1988	1.1	3.1	-2.4	4.0	3.5	3.7
1989	0.8	2.2	-2.5	2.0	1.8	2.5
1990	1.6	2.8	-1.0	0.5	-0.1	1.4

Source : OECD 1992

Entrepreneurial activity in Iceland has predominantly been in the form of investment in imported technology in the F-sector and collective forms of entrepreneurship have especially important. As the financial system in Iceland has during the post war era been dominated by state-banks and investment funds controlled by the state, state entrepreneurship has been predominant form of entrepreneurship. Investment in the F-sector has predominantly been directed through the state banks and state dominated investment funds. Furhtermore, the state is the prime motor of R&D activity in terms of expenditure as we mentioned above. In terms of performance of R&D, business enterprises perform around 10-15% of R&D activity measured in person-years (Jonsson 1991a).

The administration in Iceland is very small in absolut terms and because of great political instability in terms of duration of governments and unstable support for political parties in parliamentary and municipal elections, state entrepreneurship in Iceland is characterized by ad hoc policy formation rather than long term accumulation strategies. Basically this appears in huge investment peaks in the F-sector that last for few years while new technology is imported and new investment peaks do not appear again until after couple of decades when the technology is felt to be obsolete. Furthermore, due to low level of concentration and centralization of capital in Icleand there has not developed a powerful network of firms and/or organized interests that would be able to take over the role of forming a long term accumulation strategy (Jonsson and Jonsson). As a consequence the structural conditions for long term entrepreneurship is lacking. In 1990 the employees of the administration of the central state were 4255, excluding employees of the municipalities. The elite at policy making level of the administration was approximately 51 persons (The Economic Institute 1992 and Jonsson 1991b).

The lack of structural conditions for effective collective entrepreneurship leads to a national system of innovation in Iceland, i.e. the organization level of R&D expenditure and its objectives, which is weak in terms of R&D expenditure and in terms of being slow to shift to the new techno-eonomic paradigm of information technology and developing innovative industries on the bases of this new technology. Icelanders need to strengthen the structural base of collective entrepreneurship in Iceland if are to develop their regime of accumulation from being a technology dependent economy and become a high technolgy exporting country by means of exploiting local know-how in the F-sector and energy sector.

References

Brett, E. A., 1983, *International Money and Capitalist Crisis*, London: Heinemann.

Casson, M., 1982, *The Entrepreneur: An Economic Theory*, Oxford: Basil Blackwell.

Chesnai, F, 1988, 'Multinational enterprises and the international diffusion of technology' in G. Dosi et.al. 1988.

Cotton, J., 1992, Understanding the state in South Korea in *Comparative Political Studies* Vol. 24, No. 4: SAGE Publications.

Davis, R.D. and Ward, D.W., 1990, The Entrepreneurial State; Evidence from Taiwan in *Comparative Political Studies*, Vol 23 No. 3, 1990, Sage Publications, Inc.

Dosi, G., Freeman, C., Nelson, R., Silverberg, G., and Soete, L., 1988, *Technical Change and Economic Theory*, London: Pinter Publishers.

Economic Institute (various years) *Atvinnuvegaskyrslur*, Reykjavik: The Economic Institute.

Freeman, C., 1987, *Technology Policy and Economic Performance; Lessons from Japan*, London: Pinter Publishers.

Hodgson, G.M., 1988, *Economics and Institutions; A Manifesto for a Modern Institutional Econoomics*, London: Politiy Press.

Hoffman, K., Kaplinsky, R., 1988, *Driving Force; The Global Restructuring of Technology, Labor, and Investment in the Automobile and Components Industries*, Boulder: Westview Press.

Huang, C., 1989, The state and foreign investment; the cases of Taiwan and Singapore in *Comparative Political Studies* Vol. 22, No. 1: SAGE Publications.

Jonsson, F. and Jonsson, I., 1992, *Innri hringurinn og islensk fyrirtaeki*, Reykjavik: Felags- og hagvisindastofnun Íslands.

Jonsson, I., 1991c, 'Velferdarkerfi heimila og fyrirtaekja a Islandi i ljosi valdavaegis samfelagslegra afla', *Thjodmal, arbok um samfelagsmal 1990/91*, Reykjavik, The National Institute of Social and Economic Research.

Jonsson, I., 1992, 'Microeconomies and regimes of accumulation; an analytical framework' in Lyck, L., 1992.

Jonsson, I., 1991a, *Hegemonic Politics and Accumulation Strategies in Iceland 1944-1990. Long Waves in the World Economy, Regimes of Accumulation and Uneven Development. Small States, Microstates and Problems of World Market Adjustment*, Ph.D. Dissertation, University of Sussex.

Jonsson, I., 1989, Hegemonic Politics and Capitalist Restructuring in *Thjodmal; arbok um samfelagsmal*, Reykjavik: The National Institute of Social and Economic Research.

Jonsson, I., 1991b, 'Keynes's General Theory and Structural Competitiveness', *Thjodmal, arbok um samfelagsmal 1990/91*, Reykjavik, The National Institute of Social and Economic Research.

Jonsson, I., 1986, 'Thatcherisminn og efnahagskreppan- nyfrjalshyggjan, audvaldskreppan og hin nyja taekni' in *Rettur*, 3 and 4, Reykjavik: Rettur.

Julius, D. *Global Companies and Public Policy; The Growing Challenge of Foreign Direct Investment*, London: Pinter Publishers.

Kaplinsky, R., 1984, *Automation; The Technology and Society*, Harlow: Longmans

Lyck, L., 1992, *Nordic Arctic Research on Contemporary Arctic Problems*, Copenhagen: Nordic Arctic Researh Forum.

Nelson, R., 1988, 'Insitutions supporting technical change in the United States' in Dosi, G. et. al. 1988.

Nordic Council of Ministers (various years) *Yearbook of Nordic Statistics*, Nordic Council of Ministers: Copenhagen.

OECD (various years) *Economic Outlook*, Paris: OECD.

OECD, 1989, *Labour Force Statistics 1967-1987*, Paris: OECD.

OECD, 1992, *Historical Statistics 1960-1990*, Paris: OECD.

Rothwell, R. and Zegveld, W., 1981, *Industrial Innovation and Public Policy; Preparing for the 1980s and the 1990s*, London: Frances Pinter Ltd.

Schumpeter, J.A., 1981, *Capitalism, Socialism and Democracy*, London: George Allen & Unwin.

Yarrow, G.K., 1985, 'Welfare losses in oligopoly and monopolistic competition' in *The Journal of Industrial Economics* Vol. XXXIII, June, Oxford.

The Quiet Life of a Revolution
- Greenlandic Home Rule 1979-92[1]

Finn Breinholt Larsen

Introduction

The first of May 1979 marked the beginning of a new era in Greenland. On that date a slice of political power was cut off from the Danish government and handed over to the Greenlandic politicians. After more than two hundred years of colonial rule and an interlude of some twenty five years as an "ordinary" Danish province (*amt*) Greenland became an autonomous region within the Danish Realm.

The inception of Home Rule came after a decade of political mobilization and ethnical radicalization so when the first Home Rule government took office great expectations were attached to it by the public in Greenland. The Greenlandic politicians in charge had a lot to live up to and as could be expected not every hope was met. A survey of public attitudes conducted in 1989 showed, however, that the achievements of the first ten years of Home Rule were evaluated favorably by a majority of the electorate. [2] The drab realities of political life have obviously not worn away popular support to the political leadership.

---

[1] Besides the references cited below this paper is based on observations made during nine years as a researcher and associate professor at University of Greenland, Nuuk (1983-). An invaluable but hard to substantiate source of knowledge has been the countless number of informal discussions I have had throughout this period with Greenlandic government officers, politicians, students etc. These discussions have greatly enhanced my understanding of the vagaries of political power.

[2] 20 pct thought that the Greenlandic politicians had done a good job, 51 pct that the politicians had done a fairly good job while 16 pct gave a negative evaluation of the politicians effort. 13 pct held no opinion on the question (Henrik Skydsbjerg, unpublished data).

But no government can afford to rest on its laurels. Since the late eighties Greenland has experienced a downward economic trend and the Home Rule authorities' efforts to turn the tide has until now been in vain. An ambitious new industrial policy has been launched to reduce dependence on the fishing industry, but the outcome is still very uncertain. Furthermore, there is a growing popular discontent due to a number of cases where the new leaders have mismanaged their responsibilities. During the past several years, the Greenlandic newspapers have been covered with stories about politicians abusing public funds for personal gain. The political leaders in Greenland are in the years to come facing some tough challenges. They need to strengthen the country's economy as well as change their own image in the eyes of the people.

This article offers an analysis of Greenlandic Home Rule. What is the set-up of governance? What has been accomplished within this set-up? And what will be the main challenges in the years to come?

The Legal framework of Home Rule

The legal basis for Home Rule is the Greenland Home Rule Act (Act No. 577 of November 1978). One crucial prerequisite for the success of this act has been the gradual participation of the Greenlandic population in the political decision making process, which actually started more than 100 years ago. The first local councils with representatives from the Greenlandic population were established during the decade of 1860s. Two provincial councils were established in 1908 and the local councils were replaced by municipal councils. In connection with the decolonization, which took place after World War II, the provincial councils merged into one council and the Greenlanders obtained two seats in the Danish Parliament. The provincial councils lacked legislative power, but the Danish government was quite sensitive to the different opinions expressed in the council. A substantial change of the administrative power structure occurred in 1975, when the responsibilities of the local administrations were transferred from the state to the municipal councils (cf. Sørensen, 1983).

Table 1.	Issues listed in the Greenlandic Home Rule Act which may be assumed by the Home Rule Authorities ordered Chronological by the Year of Transfer
1979 -	Organization of Home Rule in Greenland.
-	Organization of local government.
1980 -	Direct and indirect taxes.
-	Fishing in the territory, hunting, agriculture and reindeer - breeding (partly - the rest was transferred in 1985).
-	Labor market affairs.
-	Education and cultural affairs, including vocational training (partly - the rest was transferred in 1981).
-	Social Welfare.
-	The Established Church and dissenting religious communities.
1981 -	Preservation of wildlife.
-	Country planning.
-	Legislation governing trade and competition including legislation on restaurant and hotel business, regulations concerning alcoholic beverages and regulations concerning closing hours of shops.
1985 -	Other matters relating to trade, including the State-conducted fishing and production; support and development of economic activities.
1986 -	Supply of goods.
-	Internal transport of passengers and goods.
1987 -	Rent legislation, rent support, and housing administration.
1989 -	Protection of the environment.
1992 -	Health services.

The core of the Greenland Home Rule Act is the transfer of legislative and administrative powers in particular fields to the Home Rule authority. The Home Rule Act provides for the establishment of a legislative branch, the *Landsting* , with legislative power over certain agreed upon fields of responsibility within the territory of Greenland. Furthermore, the Act also provides for the set-up of a local government, the *Landsstyre*, which has the administrative authority over these fields.

The Home Rule Act contains a list over fields of responsibility which may be transferred to the Home Rule authorities on request by the Greenlandic Government. No exact timetable for the transfer of individual fields was laid down, but all the different fields of responsibility, mentioned in the Act, were subjects for transfer during the period between May 1, 1979 and January 1, 1992 (cf. Table 1).

Fields of responsibility, not specifically mentioned in the Act, can be assumed after negotiations between the central Danish authorities and the Home Rule authorities. However, due to the Danish constitution, some fields of responsibility cannot be transferred to the Home Rule authorities. Matters affecting the foreign relations of the Realm are specifically reserved to the central Danish authorities, and the supremacy of international treaties over local powers is explicitly pronounced in the Home Rule Act. However, the Greenland authorities may comment on proposed treaties which would affect Greenland's interests, and in the commercial area, Greenland may (with the approval of the central authorities) negotiate directly with foreign governments, participate in international negotiations, and demand that Danish diplomatic missions employ officers specifically to attend to Greenland's interests in countries of special commercial importance to Greenland. Areas involving defence policy and monetary policy are also reserved to the central Danish authorities.

The Home Rule Commission, who drafted the Home Rule Act, argued that the constitution does not allow for the transfer of power to the Home Rule authorities over areas such as the police and the judiciary. Moreover, the Home Rule authorities should not be permitted to lay down the rules of fundamental principles regarding the law of persons, inheritance law, family law and property law (Betænkning 837 vol. 1, 1978: 23). However, on this point the correctness of the Commission's interpretation of the constitution has been questioned (Harhoff, 1982).

One of the most controversial issues in relation to the establishment of the Home Rule Act has been the right of ownership to mineral resources of the underground. Contrary to all other fields, where the decision making power is vested in either the Danish government or in the Home Rule authorities, decisions concerning preliminary study, prospecting and exploitation of mineral resources must be made jointly. Both parties are empowered with the right to veto any measures of fundamental nature in the area of natural resources. This should be viewed as an expression of a political compromise; the Greenlandic members of the Home Rule Commission wanted the ownership of the underground to be transferred to the people of Greenland, but this was not possible due to strong opposition from the Danish government (Foighel, 1980).

When the Danish authorities draft new laws in fields already within the competence of the Home Rule authorities, the governing rule is that such laws are invalid in Greenland. Other laws, drafted by the Danish authorities, must be made available for comments by the Home Rule authorities before they can legally be enforced in Greenland. In reality, no new Danish law will be rendered valid in Greenland without the Home Rule authorities' affirmative consent thereto (Harhoff, 1987).

A quantitative indication of the amount of competence transferred to the Home Rule authorities since 1979 is given in Figure 1 and 2. The two variables used are: The number of laws, statutes etc. promulgated, between 1975-1991, by either the Danish authorities or the Home Rule authorities and with validity in Greenland (Figure 1), and the number of public employees, divided between state, provincial council/home rule and local government employment in Greenland during the same period (Figure 2).

As shown in Figure 1, there has been a significant drop in the number of laws, statutes etc. promulgated by the Danish authorities since the implementation of the Home Rule authorities. The legislation by the Home Rule authorities has increased proportionally. However, it was not until 1990 that the laws promulgated by the Home Rule authorities exceeded the number of laws enacted by the Danish authorities.

Figure 1. Laws, statutes etc. Promulgated by the Danish Authorities and the Home Rule Authorities Respectively, 1976-1990.

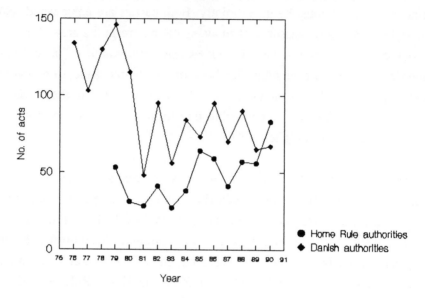

The development of the employment sector has been even more significant. In 1975, 81% of the publicly employed in Greenland worked for the Danish government. In 1990 this number had dropped to 18%. In 1992, when the Home Rule authorities assumed the responsibility of the health service sector, the number continued to decrease, and is today at approximately 5%. The majority of the 300-350 persons employed by the state are working at the police department, the courts and within the prison services. The number of public employees working at the local government increased rapidly by the mid-1970s. This growth can be attributed to the take-over of the administrative responsibilities on the local level by the municipal councils. The provincial council's administrative sector was quite modest. During the first couple of years after the implementation of the Home Rule authorities, the employment number, within this sector, was also low. But a definite increase of personnel within the Home Rule authorities occurred during the mid-1980s, as a result of the takeover of the Royal Greenland Trade Department and the other state-owned corporations.

Figure 2. Number of public employees in greenland (full-time, monthly salaried) divided between state, provincial council/home rule and local government, 1978-1990.

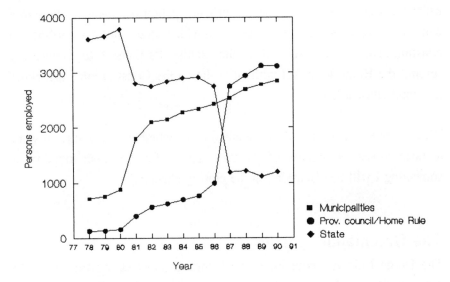

The two figures show that, within the legal framework of the Home Rule authorities, a significant transfer of the legislative and administrative powers from the Danish authorities to the Home Rule authorities has been accomplished. In reality, these figures underestimate the influence of the Home Rule authorities since all daily government functions are conducted by the Home Rule authorities. Until 1989, there was a Minister of Greenlandic Affairs in the Danish government. At that time, the transfer of tasks to the Home Rule authorities had reached such a high level that a decision to abolish the Ministry of Greenlandic Affairs was made. Instead, it was decided that the remaining tasks should be divided between other ministries. Today, there is only a small Greenlandic branch in the Prime Minister´s department.

Although formally, the Kingdom of Denmark still is a unitary state, it has, in reality, developed into something more like a federal state consisting of Denmark, the Faroe Islands and Greenland. Each constituent state has its own legislative and executive branch, while the Danish government continues to operate as a federal government. The North Atlantic constituent states have, due to their four seats (out of 179 seats) in the Danish parliament, some, although minor, influence on the decisions made by the central power. The Danish Parliament's delegation of power to the Faroe and the Greenland Home Rule authorities does not, according to constitutional law experts, allow the Danish Parliament to unilaterally rescind the Home Rule Act without violating the Constitutional principles and International law.

Whether this "federal state" is a stable one, or whether it will move toward a much looser association of the three parts of the Danish kingdom is something I will be discussing in the closing chapter.

The Greenlandic parties

The Home Rule Act constitutes the formal framework within which the political life in Greenland has developed since 1979. The Greenlandic political parties have played a central role in completing the policy goals of this Act. Back in the 1950s and 1960s a few, unsuccessful attempts were made to establish branches of Danish parties in Greenland. The creation of political parties in Greenland did not commence until the 1970s and 1980s, thus the Greenlandic party system is a fairly recent phenomenon. (For a thorough analysis of the development of the political system in Greenland see Dahl, 1985, 1986a, 1986b, 1988.) The Siumut (= Progress) party is not only the oldest of the remaining parties, it is also the first genuine party in Greenland.

The Siumut party was formed in 1977. The party was an offspring of the political movement which had, since 1969, been united under slogans such as: "The new policy", "Development in Greenland on Greenlandic terms" and "Greenlandization". The core of the movement was made up by young radical Greenlanders, who hoped to settle the score with the well established Greenlandic elite who approved of the Danish development policy for Greenland and who had a strongly co-operative attitude towards the Danish government. The principal goal of the post-war generation of Greenlandic politicians was to gain equality for the Greenlandic population within the Danish Realm with respect to standard of living, political rights etc. The criticism of this assimilationist policy was expressed in ethnical terms. The movement did not oppose the modernization as such, but was against the "Danification" of the Greenlandic society.

The movement drew its strength from the alliance between young intellectuals, who were inspired by the anti-imperialistic movement in the Western metropoles, and from the large groups of the Greenlandic population especially, the village people, who felt that their social and cultural identity was endangered by the extensive modernization program launched by the Danish Government. Symbols of ethnicity borrowed from the traditional Greenlandic hunting culture were extensively used by the movement.

"The new policy" had its break-through in 1971 when two of the leading members of the movement, Jonathan Motzfeldt and Lars Emil Johansen, were elected to the Greenlandic provincial council and a third member, Moses Olsen, was elected as one of the two Greenlandic representatives to the Danish parliament. In 1972, Jonathan Motzfeldt put the question of Home Rule on the political agenda. This took place immediately after Denmark and Greenland, but not Faroe Islands, had entered the European Economic Community (EEC). A referendum held in Denmark and Greenland disclosed that a majority of the Danish population was in favor of EEC membership. This was not, however, the case in Greenland, where the "no"-votes were predominant. The Faroe Islands' Home Rule authorities had informed the Danish government that it had no interest in becoming a member of the EEC. This was accepted by the Danish government. However, the Danish government refused to endorse the Greenlandic provincial council's request to hold a separate Greenlandic referendum on the issue of EEC membership. This refusal was based on the fact that Greenland was, constitutionally, a fully integrated part of the Danish Kingdom.

The EEC membership issue worked as a spring-board for the Siumut movement's principal political concern, the Greenlanders' urge for self-determination. The EEC became a symbol of everything that the Siumut movement opposed. The EEC was seen as a menace to the already insignificant influence the Greenlandic politicians had on issues concerning the exploitation of fishery- and mineral resources. Furthermore, the EEC membership would expand the cultural influence from Europe as well as increase the movement of foreign capital and labor into Greenland. After the inception of Home Rule a second referendum on the issue of EEC membership was held in February 1982. The result of this referendum lead to Greenland's withdrawal from the European Community, which was also supported and accepted by the Danish government. Despite this result, the main theme of the political debate continued, up until the final agreement had been signed in 1984, to center around the membership issue.

Another trademark for the Siumut movement during the 1970s was the fight against the offshore oil drilling off the West Coast of Greenland. A number of international oil companies had been issued concessions to an area of the continental shelf of Greenland. But after a few fruitless drilling attempts had been conducted in 1974-75, these concessions were immediately rescinded. This was yet another matter with symbolic dimensions: The fight against the multinational oil companies' activities in Greenland symbolized the fight against the Western imperialism and the right of self-determination of Greenland's natural resources (Davies et al., 1984). In 1975, when the Home Rule Commission began its work, Lars Emil Johansen, who was Siumut's representative on the commission, demanded that the Greenlandic population's ownership rights to Greenland's underground resources should be recognized. As mentioned above, this demand was not met in its entirety. Nevertheless, it played a significant part in the formation of the political consciousness, especially among the younger generation in Greenland and it also became an important political mobilization factor.

The leaders of the Siumut movement had, long before the party's creation in 1977, won a place among the Greenlandic political elite. These leaders were argumentative, energetic and intelligent politicians who distinguished themselves in the eyes of the Greenlandic population, and gained significant impact in the political debate not only in Greenland but in the Danish media as well.

The ideology of the party rested on moderate socialism coupled with nationalism. This ideology, in addition to, the leaders' well-developed sense for the political game, made the Siumut party the number one party. In fact, at the *Landsting* election in 1979, the party secured no less than 13 seats out of 21 possible. Additionally, it occupied all positions of the Home Rule government and has been in power ever since. Jonathan Motzfeldt, was premier of the Home Rule government between 1979-1991 at which time he was replaced by Lars Emil Johansen. The representative for the party in the Danish parliament worked, for many years, closely with the Socialist Peoples' Party but has in recent years been linked to the Social Democratic group.

As soon as the Siumut party became a permanent party, others followed suit. In 1977 the Atassut (= Mutual Connection) party was created as a loosely organized group of provincial council members. In 1981, the party became an actual constituent organization. The party originated as a reaction to the Siumut party's provocative style towards the Danish government. Thus, the party was prone to the thinkings of the old political elite which favored a smoother relationship with Denmark. The party supported the idea of the Home Rule, but wanted to avoid that the autonomy movement became an actual separatist movement. Moreover, the Atassut party was a strong supporter of continued EEC membership and in favor of the exploitation of oil- and mineral resources. Ideologically Atassut has always placed itself to the right of the Siumut party. Initially, the representative for the party in the Danish parliament was affiliated with the Social Democratic group, but is today aligned with the liberal party, Venstre. Even though the Atassut party has, periodically, been the largest party in Greenland, it has never held a position in the Home Rule government.

In recent years, the Atassut party has experienced competition from the new center-conservative parties: Issittup Partii-a, which was established in 1986, and Akulliit Partiaat which was created in 1991. Both parties have achieved some representation in the Home Rule parliament, mainly at the expense of the Atassut party. All three parties want to promote the private sector and limit the scope of the public sector. However, Issitup Partii-a has a more nationalistic party line than the Atassut party and has, therefore, been able to appeal to the non-socialist voters, with feelings of nationalism.

Siumut's political leaders have, from the very beginning, had a rather pragmatic style. They preferred compromise rather than political deadlock and intransigence. One case, which clearly reflects this attitude, is the question relating to the ownership of the underground resources, where the Siumut leadership was willing to settle for less than a 100 percent transfer. However, the left-wing members of the party disagreed with the rest of the party on this issue and as a consequence created the left-wing group, Inuit Ataqatigiit, in 1978. This group was later transformed into a regular party.

The Inuit Ataqatigiit party rejected the Home Rule Act because it did not recognize the Greenlandic populations' ownership rights to the underground resources. The party ran, unsuccessfully, for the election to the Home Rule parliament in 1979. But the party has, however, since 1983 been represented in the *Landsting*. And at the latest election, the party received approximately 20% of the votes and is today the third largest party in Greenland. Early on, the party proclaimed to be a marxist-leninist party and used anti-imperialistic and anti-capitalistic slogans. Furthermore, it emphasized, more than any other party, the Greenlandic ethnicity. The party has, since it was established, moved closer to the center and may today be characterized as being a left-wing Social Democratic party with a strong twist of nationalism. The Inuit Ataqatigiit party has during three terms worked as a coalition partner to the Siumut party in the Home Rule government and a considerable alignment of the two parties' political positions has occurred in recent years. Rumors have it that in 1992, the two parties were negotiating the possibility of uniting.

The leadership of the Inuit Ataqatigiit party has, since its inception, been dominated by the outstanding political personality, Aqqaluk Lynge. It is, partly, due to his leadership abilities that the party has been able to afford such a pragmatic turn without loosing the radicals' vote in Greenland.

A significant characteristic of the party system in Greenland is that the establishment of the system occurred in conjunction with the changes of Greenland's constitutional position. During the first couple of years, the fundamental dimension of conflict within the party system centered around the parties' different attitudes to the question of Greenland's future relationship with Denmark. In light of this conflict, it is quite understandable that the attempt made by the labor movement to form a labor party, the Sulisartut Partiiat, did not succeed. The clash between the workers and the management was completely overshadowed by the issue of ethnicity and the party was, a few years later, absorbed by the Siumut party. Although, the dominating party of the labor movement is the Siumut party, the Inuit Ataqatigiit party also has a significant influence on the movement.

It is important to note that the conflicting interest regarding Greenland's future relationship with Denmark was, primarily, not a matter between Denmark and Greenland but, rather, an internal conflict of interest within the Greenlandic population. This conflict is closely linked to the feelings of ethnical and cultural identity. Although, this search for national identity started already in the beginning of this century (cf. Thuesen, 1988), it did not reach political dimensions until the 1960s. What kind of political nation should Greenland develop into? Atassut's answers was, that Greenland should be a political nation with close ties to Denmark, but at the same time, have a special Greenlandic identity. The Siumut party strived for a political nation with a looser connection to Denmark but with closer ties to the rest of the Inuit world. Inuit Ataqatigiit wanted Greenland to be an Inuit nation which, when possible, should cut all ties to Denmark and enter into a close cooperation with the other Inuit nations.

A political survey, conducted in 1984, illustrates these different attitudes. The question posed to the voters was, whether Greenland should prioritized the cooperation with the Nordic countries over the Inuit population or vice

181

versa or if the cooperation should be put on a par (table 2). Regardless of party affiliation, only a few voters believed that the cooperation with the Nordic countries should be of

primary concern. Hence, this result lends itself to the interpretation that, even among the part of the population in favor of a close link to Denmark, there is no desire of an exclusive Nordic identity. However, there is an excess of voters the Inuit-cooperation should be on the top. The majority of the voters within the Inuit Ataqatigiit argues that the Inuit cooperation ought to be prioritized.

The respondents were also asked to whom they felt a greater closeness - the Greenlandic population or the Danish population, or if they felt an equal attachment to both groups.

Table 2. Greenlandic Voter's Attitudes Regarding External and Internal Association (only persons born in Greenland are included)

	Atassut	Siumut	Inuit Ataqa- tigiit
Cooperation with the rest of the Inuit people and people in other Nordic countries			
Cooperation with the Nordic countries should be prioritized	9	9	4
Cooperation with the Inuit population should be prioritized	14	27	52
Cooperation in both directions should be prioritized	49	38	26
Don't know/unanswered	28	27	18
Total	100	100	100
Association to the Greenlanders and to the Danes in Greenland			
Feel more closely connected to the Danish population	0	0	1
Feel more closely connected to the Greenlandic population	60	77	89
Feel equally as close to both groups	37	20	9
Don't know/unanswered	3	2	1
Total	100	100	100

Source: Finn Breinholt Larsen & Per Langgård, unpublished data.

The distribution of answers follow, with some modifications, the same pattern as the previous question. By and large, no one among the Greenlandic electorate feel a greater attachment to the Danes. The majority of all the persons asked, answered that they are more linked to the Greenlanders. But the number of people who feels attached to both population groups increases significantly from the left to the right side of the political spectrum.

Even though none of the Greenlandic parties has questioned the legitimacy of a Greenlandic nation, the fight over Greenland's association to Denmark has, nevertheless, been an issue which has greatly divided the Greenlandic population. The demarcation, has, especially, been between the Atassut wing and the Siumut-Inuit Ataqatigiit wing. The national "identity conflict" has, for a while, separated the society into two irreconcilable camps. The fight has divided many a family and village in two hostile groups (cf. Dahl, 1985).

Since the Home Rule authorities assumed the responsibility for many different areas of the society, the identity conflict has to some extent been pushed aside to leave room for more pragmatic political solutions. This blur of the dividing line in Greenlandic politics has also paved the way for the re-alignment of the electorate. Table 3 shows the party affiliation at the past five *Landstings* elections. Today, there are 5 parties in the Home Rule parliament, as opposed to only 3 in 1979.

Table 3. Distribution of Votes at the *Landsting* Elections 1979-1991.

	1979	1983	1984	1987	1991
	PCT	PCT	PCT	PCT	PCT
Siumut	46.1	42.3	44.1	39.8	37.3
Atassut	41.7	46.6	43.8	40.1	30.1
Inuit Ataqatigiit	4.4	10.6	12.1	15.3	19.4
Sulisartut Partiiat	5.6	-	-	-	-
Issitup Partiiat	-	-	-	4.4	2.8
Akulliit Partiiat	-	-	-	-	9.5
Independent candidates	2.2	0.5	-	0.4	0.9
Total	100.0	100.0	100.0	100.0	100.0

Source: Grønland, 1990 (including figures for the 1991 election)

As already mentioned, the Siumut party has managed to stay in power ever since the inception of the Home Rule. It was, however, only between 1979-1983, that the party reached absolute majority in the *Landsting*. During 1983-84, the Siumut party formed a minority government which was later overthrown by the Inuit Ataqatigiit and the Atassut parties. In replacement, a majority government, including both Siumut and the Inuit Ataqatigiit parties, was established. This coalition was, however, under dramatic circumstances, dismantled in 1987 only to be re-created after the *Landsting* election. This time, the cooperation lasted until 1989, when the Inuit Ataqatigiit once again was forced out of the government. The Siumut party, with the political support of the Atassut party, continued to rule as a minority government. After the election in 1991, the coalition between the Siumut and the Inuit Atagatigiit was reinstated. Moreover, the Siumut politicians hold the majority of the mayoral positions in Greenland's 18 municipalities.

Compared to most post-colonial societies, the political system in Greenland shows a high degree of stability and democracy. This, however, does not mean that Greenland has managed to avoid political power struggles and political scandals. Immediately after the *Landsting* election in 1987, a faction, consisting of high powered members of the Siumut party tried, under Lars Emil Johansen's leadership, to remove Jonathan Motzfeldt from his position as head of government. The fact that the group decided to strike at a time when Motzfeldt was away from Nuuk added a flavor of coup d'etat to the whole episode. However, the attempt to jockey Jonathan Motzfeldt out of his post failed, and Motzfeldt continued to be in charge of both the government and the party (Larsen, 1988).

At the Siumut's party convention in 1990, Lars Emil Johansen once again, challenged Jonathan Motzfeldt by running for the party chairmanship. Lars Emil Johansen did, this time, bring home the victory. In 1991, on the night before the election to the Home Rule Parliament, Lars Emil Johansen, rather unexpectedly declared that he would run for the position as premier of the Greenland Home Rule government. The Siumut faction, led by Lars Emil Johansen had strengthened its position during this election and a few weeks later, after intense lobbying activities, Lars Emil Johansen assumed the position as the head of government.

The power struggle between the two leaders of the Siumut party was not only a matter of personal differences. Lars Emil Johansen represents a faction of the Siumut party who wants to dissociate itself from that type of abuse of office that has been brought to light through several public scandals during the recent years, and which also has diminished the peoples' confidence in the politicians. Motzfeldt's excessive alcohol consumption and the irrational decisions that sometimes were made as a result hereof also contributed to his downfall. The *Landsting* election in 1991 was provoked by the news media's and the opposition's criticism of the Home Rule authorities' abuse of its expense account. The remaining politician in the formerly criticized group of the Home Rule government, was due to a second incident of mismanagement, which this time concerned the pilot program for farming of sea trout, forced out of the government in 1992.

This, at times, rather turbulent political life has rubbed off on the bureaucracy. Lay-offs of Government officials and managers of publicly owned companies have been seen as a recurrent theme.

Achievements and challenges ahead

In the following, I will shortly discuss some of the issues which have been in the limelight since the commencement of the Home Rule, and which will continue to be a challenge for the Greenlandic politicians in the years to come.

The economy

One of the most important consequences of the Home Rule has been the transfer of the responsibility of the economy onto the Greenlandic politicians, which has greatly strengthened their potential for influencing the societal development. But, this also meant that Greenland's economic dependency on Denmark was made more visible. With almost every area of responsibility that had been transferred to the Home Rule followed a yearly block grant. Without these block grants the standard of living in Greenland would have gone down dramatically.

Figure 3 indicates the share of the public spending in Greenland, between 1979 and 1990, which have been financed by the Danish government, both directly and through block grants.

Figure 3. The Overall Public Spending in Greenland 1979-90 and the Danish Share Thereof.

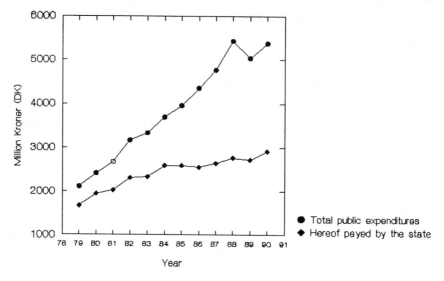

The Danish government paid in 1979 almost 80% of the public spending in Greenland. The share has since then been considerably reduced. This cut is a consequence of the fact that the Governmental spending in fixed prices, to a great extent, have been constant over the entire period, while the total of the public spending has increased significantly. The additional spending have been financed through a radical increase of taxes. The Danish government is still financing more than 50% of the public spending in Greenland and it does not look like this share will decrease in the nearest future.

The national purse experienced a serious liquidity crisis in 1987 due to bad budgetary management. As a result, stricter spending policies were implemented, which in turn has slowed down the enterprising activities and led to an increase of unemployment and bankruptcy. In addition, a number of municipalities have had problems controlling the economy which has resulted in a worsening of the economic crisis. To this must be added the failing revenues from fishery as well as the closing of the lead-zinc-mine in Maarmorilik in 1990. Real wages have decreased in almost every occupational field since the beginning of 1980s, and this trend can be expected to continue in the years ahead.

Greenland's continued economic dependency on Denmark moves the dream of an independent Greenland far into the future and is, thus, stabilizing the constitutional arrangement laid down in the Home Rule Act. The Danish government's spending in Greenland amounts to 2.9 billion Danish kroner (app. 580 million Canadian $) yearly, thereof is approx. 2.7 billion Danish kroner spent on block grants. The overall subsidy from the Danish government totals approximately 52.000 Danish kroner (app. 14.000 Canadian $) per capita in Greenland.

Industrial policy

The desire for greater economic and political freedom has, since the inception of the Home Rule, made the industrial policy one of the most important questions for the Greenlandic Home Rule authorities.

The majority of the Greenlandic trade and industry was, before 1979, owned by the Danish state. When the Home Rule authorities assumed the responsibility for these, previously state run enterprises, it also took over a tradition of public planning in areas of production and commerce. Initially, the moderate socialist Home Rule government was confident that the industrial development would boost through the use of central planning. The main priority was given to the fishing industry. Attempts to make the Greenlandic fishing industry more efficient and modern were made through extensive public investments. This, however, ended in a number of disastrous miscalculations regarding both trawlers and fishing plants. This catastrophe caused the Home Rule authorities to realize that a change of the economic policy was necessary. Since, the end of 1980s, a great effort has been made to run the publicly owned enterprises as commercial companies.

Royal Greenland, the production and export business, owned by the Home Rule authorities has been converted into a publicly owned corporation. After a record high deficit in 1987, the management has succeeded in reducing the overhead costs and thereby, managed to turn failure into success. From receiving millions every year in subsidy, the enterprise, today, shows a surplus even though this surplus still is far from satisfactorily (Royal Greenland, 1992). Further improvements of the result have been prevented due to the falling prices on shrimp which, is Royal Greenland's chief exporting product. The cod fishery has also been subject to miscalculation, during the last years, due to demise of the fishing stock. An actual privatization of the Royal Greenland is not yet on the agenda, since that may result in layoffs and closings of factories, to an extent, which would be political unacceptable.

Besides public fishing there has, for years, been a lucrative private shrimp fishery carried out in Greenland. A too large trawler fleet and falling prices on the world market have, however, caused the private fishing enterprise to experience great economic difficulties during the last couple of years. The Home Rule authorities have taken some initiatives to reduce the surplus capacity by promoting enterprise mergers and thereby improve the profitability of the private trawler fleet (Fiskeristrukturudvalget, 1989).

The Home Rule authorities' industrial policy has, from the beginning, focused on the utilization of the living resources. The Home Rule authorities hoped to create a sound economy by taking necessary steps to develop and modernize the fishing industry, as well as improve the marketing and processing of these products. Since the economic backlash, in the end of the 1980s, it has been quite obvious that Greenland needed to put greater emphasis on other occupational sectors, and thereby create a more differentiated business structure in order to survive economically. The fishing stock in the Arctic waters is, basically, instable and vulnerable. The quantity of cod fish, along the Greenlandic coast, is fluctuating (today, they are almost extinct), and it has, during the last years, been indications that point to depletion of the shrimp stock. This situation, combined with the heavy fluctuation of the market price of these products, makes it very risky to base the economic development strategy exclusively on fishery resources.

To combat this problem, the Home Rule government launched a new industrial policy in 1989. The purpose of this new policy was to stimulate mineral resource development, and to promote the development of other areas of the industrial sector, as well as to encourage tourism to the island (Råstofforvaltningen, 1990). The economical conditions, for the extraction of raw materials, have been made more favorable in order to increase the prospecting activities, and thus enhance the chances for commercial discoveries, which in turn would entice foreign corporations to Greenland. An assessment of the effects of this policy is still premature. No extraction of either minerals or oil in Greenland takes place today. Furthermore, the previous years of prospecting activities have not resulted in the discovery of commercial deposits.

There is, however, no doubt that Greenland, with its breathtaking sceneries, has great potentials for "exotic" tourism. The goal is to raise the yearly number of tourists, from the now, 5-7000 to 35.000 within the next 10-15 years (Landstinget, 1990). The infra-structure and the hotel capacity have, during the last couple of years, expanded and the Home Rule authorities support the development of private tourist operation by granting subsidies to advertising campaigns etc.

In addition, the Home Rule authorities has granted money to the establishment of import substituting enterprises (i.e. soap production, plastic production), as well as, to the development of "non-traditional" export enterprises, such as, the production of Santa Claus products.

The Home Rule authorities' initiatives, under the new industrial policy, cannot hide the fact that it is difficult to run commercial business operations in Greenland, due to the geographical conditions, the distance to the export markets and the populations relatively low educational level, etc. This, makes it particularly arduous to attract foreign investors. It should also be remembered that the Greenlandic people lack the tradition of enterprise skills. Therefore, it is apparent that the dominating business in Greenland will, for many years, continue to be the fishing industry.

Greenlandization

Greenlandization, has been one of the most common words used in the political debate in Greenland, both before and after the inception of the Home Rule. The word "greenlandization" implies both, that the greenlandic population should replace the Danish guest workers in all areas possible, and that the Greenlandic culture and mentality should influence the societal development of Greenland.

One of the first laws passed, by the Greenland Home Rule parliament was a law about the regulation of the employment recruitment in Greenland (Act No 1 of March 1980). This law gives, permanent Greenlandic residents, preference to jobs within the categories of unskilled labor, office personnel, semen and sea officers and skilled craftsmen. The objective of the law was to prevent excessive import of Danish workers (Danes with a permanent residence in Greenland are not affected by the law). It has been a political desire to guarantee that other occupations within the public institutions should, to the extent possible, be filled by Greenlanders. 2/3 of all wage earners are employed within the public sector, and the public employers have, therefore, had a great opportunity to, more directly, affect the employment policy. The share of local Greenlandic workers, (monthly salaried full-time employees) within the public sector has risen from 56% in 1978 to 71% in 1990 (cf. figure 4). Figures for the hourly employed are not available, but the number is, without any doubt, higher within this area.

It is, of course, difficult to document the extent to which the Greenlandic culture and mentality have influenced the societal development after the inception of the Home Rule. All the members of the *Landsting* and the Home Rule government are native Greenlanders. 17 out of the 18 mayors are also native Greenlanders and there are only a very few members of the local councils who are not native. The participation of the Danes in the political life in Greenland is nearly invisible, even though, the Danish vote constitutes approximately 20% of the electorate. On the other hand, the Danes still occupy the majority of the leading positions within the public administrations and thus, play an important role in the political decision making process. Since 1979 a large number of Danish civil servants have been transferred from Danish government agencies to

Figure 4. The Share of full-time, monthly Salaried Greenlanders within the Public Sector, 1978-1990.

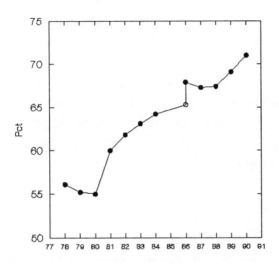

The Home Rule administration as the Home Rule assumed responsibility in their respective spheres. This may be an important reason why Home Rule in many ways has seemed to imitate the practices of the former colonial masters. The Greenlandic elite, too, has often been accused of having been influenced by the "Danish mentality" (Hansen, 1992). However, due to a high turnover rate among Danish employees in Greenland the 'old guard' has to a great extent been replaced by a new staff. This has paved the way for non-conventional solutions in many fields. Whether these should be characterized as 'Greenlandic' or not is ipso facto very much a matter of opinion.

Another aspect of Greenlandization is the effort to strengthen the Greenlandic language. The Home Rule Act states: "The Greenlandic language is the primary language. The Danish language shall be thought thoroughly." The Danish language has, in reality, been the dominating language in the schools in Greenland. In 1980, only 40% of the teachers on the teaching staff in the elementary schools were able to teach in the Greenlandic language, as opposed to 60% in 1990 (Grønland, 1990). Starting in 1994, the integration of Greenlandic and Danish school children will take place, which means that, after this date, the Danish pupils will receive part of their education in Greenlandic. Danish, continues to be the leading language in the higher educational institutions, but the communication language in the Home Rule parliament is Greenlandic with simultaneous translation into Danish. The Greenlandic language is spoken in 2/3 of all the talk shows on the radio (2500 hours/year out of 3800 hours/year), while the TV programs are dominated by the Danish language closely followed by the English speaking programs with Danish subtitle. The production of TV programs, produced in the Greenlandic language has a high political priority with the goal to fill 10% of the broadcasting time with Greenlandic programs (Ibid.).

Regional development

Traditionally, the Greenlandic population has been spread out along the vast coastline of Greenland. Through the Danish modernizing program, the Danish Government tried to move the population to certain parts of the country, mainly to the towns and the larger villages in the central and southern parts of West Greenland. As a consequence, the population figure decreased in the villages during the 1960s and 1970s, and quite many villages were completely deserted. This concentration policy created a lot of anguish among a large portion of the Greenlandic population. It has, therefore, been an important objective for the Greenlandic political parties to restore the living conditions in the villages and the outer districts.

As shown in Table 4., a continued decline of the population figure in the villages has come to a halt. Since 1980, both the number of villages and the number of village inhabitants has been unchanged. Since the inauguration of the Home Rule, electricity and water plants have been built in a great many villages. Moreover, helicopter platforms have been established in many of the villages which cannot be reached by boat during the winter periods. The Home Rule authorities are funding air routes to these sparsely populated destinations, and hereby enhancing the level of service of food supply etc. In 1985, an ambitious modernization program for the villages was approved by the Home Rule parliament. Today, as part of this program, small industrial plants, for fish- and meat preparation, are being built in several small villages. The purpose for building these plants is to create greater income opportunities for hunters and fishermen, as well as, to establish additional work places, especially for the female population.

The economic crisis Greenland is facing, has lead to a much more critical assessment of the public investment projects, thereby, handicapping the small communities' position in the competition for public funds. The good faith intentions to modernize the village communities are thus, in conflict with the Home Rule authorities' more important endeavors to rationalize and effectuate the Greenlandic society. This is, however, a sensitive political issue, since the improvement of the villages' conditions has been one of the Home Rule authorities' fundamental political goals. Only one party, the Akulliit Partiiat, has openly approved of the dismantling of villages with poor development potentials.

Table 4. The Number of Towns, Villages and Sheep-farming Places in Greenland, 1960-1991. The Village Population's Share of Total Population.

	1960	1970	1980	1990	1991
Towns	17	17	17	17	17
Villages	110	83	67	68	70
Sheep-farming places	23	40	33	38	38
The village populations share of total population	39%	25%	20%	18%	18%

Source: Grønland, 1990 (including population figures for 1991)

Jonathan Motzfeldt, the Siumut party, has approached the delicate matter by stating that, if one chooses to remain in a village, s/he also has to accept a lower standard of living than is found in the towns. The problems surrounding the villages have also been less visible in the administration of the Home Rule authorities since the close down in 1990 of the special department for village and outer district affairs (established in 1979), and the subsequent distribution of tasks between several different departments.

A report prepared by the officials of the Home Rule authorities' administration instigated a stormy debate because it seemed to, on many issues, follow the denounced Danish developmental policy. The report stated that future business operations ought to be located in towns where electricity and water could be produced at a cheaper price. This, actually, corresponds to the real development, since 80% of the total population growth during the last 10 years has been concentrated to the four principal towns in West Greenland (Nuuk, Sisimiut, Ilulissat and Qaqortoq). The tendency to population density on the regional level has not been broken by the Home Rule authorities and will most likely continue in the future.

Parallel to the population density in the larger towns, a differentiation of life style, as well as, of the material conditions of living has taken place. While, people in the small communities earns a living through a mixture of subsistence and wage work (cf. Larsen, 1987), the support of most families in the larger towns consist of wage work. The urbanization has created a, rather, skewed income distribution. A new middle class has emerged, whose life conditions are radically different from those of the fishermen and hunters.

International Relations

As previously mentioned, the Home Rule authorities has no competence to unilaterally enter into an agreement with foreign governments. Nevertheless, the Greenlandic politicians have managed to influence the various international relations Greenland is participating in. An illustration of this Greenlandic influence can be seen in its withdrawal from the European Economic Community. The Home Rule authorities, has through the Inuit Circumpolar Conference (ICC), strengthened its relations to the rest of the

Inuit world. Hans Pavia Rosing, the Siumut party's present member to the Danish parliament, was the first president of the ICC. The ICC's main significance for Greenland has centered on cultural matters. The Home Rule authorities are also participating in the tasks of the Nordic Council of ministers, both at the minister level and at the level of the officials. The Greenland Home Rule parliament has two representatives in the Danish delegation to the Nordic Council. During the 1980s, direct air routes from Greenland to the Arctic Canada (Eqaluit) and to Iceland have been established. This has made the dealings with the neighboring countries to the East and to the West much easier.

Greenland's most important foreign policy interest lies in the fishery. Greenland has representatives in the Danish delegation to the multilateral organizations which regulates the fishing operations in the North Atlantic Ocean (Northwest Atlantic Fisheries Organization, Northeast Atlantic Fisheries Organization and the North Atlantic Salmon Conservation Organization) as well as in the International Whaling Commission. Greenland is also a member of the newly founded North Atlantic Marine Mammal Commission. Furthermore, Greenland participates, to a high degree, in a bilateral fishing cooperation together with other fishing nations.

In conjunction with Greenland's withdrawal from the EEC in 1985, it gained recognition as being one of the overseas countries and territories (OLT-status), which was connected to the European Community. The agreement with the European Community allows Greenland to sell its fish products free of duty to countries within the Community, and in return, the EEC receives some fishing rights in the Greenlandic fishing zone. According to the present fishing agreement, which spans over a 5 year period, starting January 1, 1990, the EEC pays approximately 270 million Danish kroner (app. 54 million Canadian $) per year to the Home Rule authorities for the privilege to catch a fixed amount of fish (Gennemførelsesprotokol, 1989). This amount is to be paid even if the right never is utilized. Since, the Greenlandic fishermen, according to the agreement, has priority to fish within the Total allowable Catch (TAC), determined by the fishery biologists, the agreement has in, actuality, not

197

afforded the EEC fishermen with a whole lot of fish. This fishing agreement has, therefore, been especially favorable to Greenland, and it will, without doubt, require some rather crafty negotiation skills to achieve a similar agreement in 1995.

The renewal of the fishing agreement with the Common Market is bound to breathe new life into the debate on Greenland's association to the EEC. By then, the majority of the European countries have joined the EEC, or will at least be close to joining. At the same time, the political and the economical cooperation will have expanded considerably. This can make it, rather, complicated for Greenland if it chooses not to participate which, on the other hand, will present the "Europeans" in the Greenlandic parties with valuable reasons for a closer affiliation with the European Community.

Social Problems

The core in the Scandinavian social welfare model is based on the public's responsibility to secure all its citizens a minimal amount of material welfare. By copying the Danish welfare state model on to the Greenlandic society, the Danish politicians tried, in the 1960s and 1970s, to put the Greenlandic population on equal footing with the Danes, living in Denmark. Schools, housing, and hospitals were built, as well as, the social service system and educational system were expanded. The Home Rule authorities has assumed the responsibility of the distribution of the various welfare benefits and has also made efforts to further the development, once started, by the Danish authorities. Unfortunately, there continue to be short-comings within numerous areas.

The idea of introducing a public unemployment insurance was abandoned in 1987, because it was found to be way too expensive. Unemployment has been on the rise since the beginning of the 1980s and is, especially, high among the unskilled workers. The unemployed can only receive ordinary social benefits which, for families hit by unemployment, have resulted in a significant decline in income, hence, a lower standard of living.

In addition, it has been impossible to reduce the shortage of housing despite, an extensive building plan both before and after the inception of the Home Rule. The waiting lists, for an apartment in a public owned apartment complex, which in many instances is the only affordable, are extremely long. The wait can be up to several years, which has caused many to live together in apartments, already too crowded.

Despite the geographical conditions, the public service support has, within many other sectors, reached a high standard. An example is the health service sector. Greenland has been able to introduce a look-alike version of the Danish welfare state. But, like many other affluent countries, the expansion of public benefits has not, necessarily, increased the quality of living, neither has it reduced the social problems.

The social malaise in Greenland could be characterized by the following catchwords: alcohol abuse, violence, sexual crimes, suicide, accidents, absenteeism. The various mischiefs are often connected to, and enhanced by, alcohol consumption. The alcohol policy is one of the most controversial subjects amongst the population in Greenland. The attitudes of the various Greenlandic governments towards the use of alcohol have, except from a short period of liquor rationing (1982-83), been liberal. The attitudes have been that the drinking habits and the alcohol consumption could only change through the influence of consumer campaigns but not by restricting the sale of liquor.

The effect of this alcohol policy has, until today, been insignificant. There has, however, been a decline of the average alcohol consumption per capita during the last years, but this is more likely caused by a general drop in the economy than a case of alcohol policy. The number of alcohol related crimes, accidents and suicides is still, frighteningly, high. Greenland has one of the highest suicide rates in the world, and the murder rate is 20 times higher than that of Denmark. The explanation for these, depressing numbers is the way of drinking, associated with a number of cultural and social factors (Larsen, 1992). The pattern is well known from other countries in the Fourth world which have undergone the same rapid change from being a traditional country to becoming a modern society.

Generally, one must hold that the Home Rule authorities has not yet been able to reduce the social problems which cause the quality of life to deteriorate. This, however, does not mean that nothing is being done to combat these problems. In 1986, the Danish state did in cooperation with the Home Rule authorities, establish a prevention council (Paarisa) which, inter alia, has run educational campaigns against AIDS, alcohol and sniffing. Additionally, most municipalities have employed a prevention counselor, with the function of running educational programs at the schools and in the neighborhoods. The members of the present Home Rule government have tried to become role models for other citizens by publicly speaking about their own battles with alcohol and how they have managed to overcome these problems.

Education

The most frustrating problem for the Greenlandic population, as a consequence of the Danish modernization program in Greenland, has been the high number of Danish guest workers, which increased the total share of Danes in Greenland from 2% of the population in 1945 to almost 20% in 1975. The creation and the operation of a modern society required educational qualifications, which the Greenlandic people lacked and could not acquire in a short period of time. The Greenlanders were, therefore, pushed to the side line and could, as bystanders, only watch what was happening to their country. This, of course, created feelings of powerlessness and resentment. Adding insult to injury, the authorities enacted a law which legitimized wage discrimination. All of this laid the foundation for the ethnical radicalization during the 1970s.

Long before the creation of the Home Rule, the Danish Government had made great attempts to educate young Greenlanders both practically and theoretically. These efforts were continued by the Home Rule authorities when they assumed the control of the educational sector in 1980. Today, the

Greenlandic educational system is so extensive that most educations with relevance for the Greenlandic society can be carried out in Greenland.

Initially, the goal was to educate as many young Greenlanders as possible so that the Greenlandic work force could replace the Danes. Sadly enough, the quality of the education has proved to be below par. The urge to quickly accomplish the Greenlandization is thus, conflicting with the desire to achieve economic growth. Today, inefficiency of the educational system is the foremost growth restricting factor of the Greenlandic society. As opposed to the geographical and climatic conditions, the educational factor is one that can be influenced and resolved by taking the right measures. A high educational level may, to some extent compensate for the lack of rich natural resources and thus, form a basis for economic development. The Greenlandic authorities are facing a tremendous challenge in their effort to raise the level of the educational system, beginning at the elementary school level.

Conclusion

It is no exaggeration to describe the inception of the Home Rule as a revolution. The Home Rule implied a radical change of the distribution of powers. Equally as important was the change of the mental state of the Greenlanders which happened concurrently. From having been an ethnical minority on the outskirts of Denmark, the Greenlanders became a self-governing nation within a, de facto, federal government. This has created a new consciousness and proudness amongst the Greenlandic population, and have operated as the driving force in forming the many reforms the society has undergone since the origin of the Home Rule.

The results the Home Rule authorities have accomplished, must be described as impressive. During the 1960s and 1970s, the Greenlandic politicians' main criticism of the Danish Government was directed towards the Danish modernization policy in Greenland. The politicians argued that the implementation of the program was happening much too fast. In light of this criticism, it is both ironical and interesting to note that the speed has not slowed down since the Home Rule authorities took over. The Home Rule authorities has fiercely continued on with the Danish modernization project. The major difference is, however, that now, the responsibility lies in the hands of the Greenlandic politicians. The distance between the governing and the governed has decreased culturally as well as geographically. This has contributed to a much higher degree of legitimacy of the political regime.

The continuation of the Danish Government's modernization project can be attributed to the fact that the Greenlandic politicians are under a lot of pressure to make improvements in almost all areas of the society. The Danish developmental policy instilled, in the Greenlandic population, the hope that the living standard should reach a level comparable to that of Denmark. This provoked an authentic revolution of rising expectations. Denmark had, for many years, functioned as a role model for the Greenlanders in their quest to achieve the same standard as in Denmark in sectors such as the health service and the social services etc. The ethnical consciousness and the partial independence from Denmark made the comparisons to the Danish conditions less relevant. Instead, the expression "to elevate the standards to the international level" has been more commonly used during the last years. Behind this, somewhat vague expression, lies an ambition that Greenland technically and economically etc., shall reach the same level as the richest countries in the Western hemisphere.

In lieu of the following factors: the Greenlandic peoples' educational level continues to be low, the country's natural- and living- resources are hard to get to or are fluctuating, the far distance to the export markets and the short supply of both capital and organizational resources; one has to ask if the level of ambition is not set too high. It is also characteristic that the optimism, which helped to influence the developmental changes in the beginning of the Home Rule, was seriously undermined in 1987, when the country's economical situation drastically deteriorated. However, it is very difficult to adjust the populations' expectations downward. It should be remembered that the standard of living on the social and geographical outskirts of the Greenlandic society continues to be low. It is, therefore, quite possible that the distribution conflicts will be intensified in the years ahead: the distribution between rich and poor, between towns and villages, and between the municipalities in the mid- and the southwest of Greenland and the municipalities in the periphery, etc.

Consequently, the political life will to a large degree focus on Greenland's internal situation. The framework for the political life, the Home Rule Act, will most likely not be put on the agenda. A significant stand point on this issue came in 1991, when Hans Pavia Rosing, the parliament representative for the Siumut party, declared to the Danish public that Greenland within the next 10-20 years was seeking a much more relaxed association with Denmark. This statement was quickly refuted by the premier of the Home Rule government, Lars Emil Johansen. The Greenlandic General Executive did not want to create any uneasiness regarding the issues of Greenland's affiliation with Denmark. In a recent interview, Lars Emil Johansen described the relationship to Denmark as such: "During the years I was a member of the Danish Parliament (1973-1979), everything looked differently. Not a single one of us expected that it all would happen so quickly. We have, in every area, exceeded the expectations that was proposed in the report by the Home Rule commission. Today, 90% of all our affairs are regulated by the Greenlandic authorities. But we are still facing tensions. I was, before the inception of the Home Rule, very critical because all the decisions concerning the society of Greenland was made 4000 km. away. The cooperation has, however, worked pleasingly. The Danish parliament exhibits a considerable amount of fairness which we would like to return. One should not discard all the good things that have been reached throughout centuries. If, the Danish Government would interfere and hinder our political regulation of the society, then, it would be a complete different matter. But this is not the case at all. It is not just a matter of the three billion Danish kroner which we receive from the Danish State each year. No, this is much more deeply rooted. We have several hundred years of co-existence, family ties, friendship, and human relationships. - Even if Greenland and the Faroe Islands were economically independent, we would still have significant issues that needed to be coordinated. *(Interviewer) Someone mentioned something about a federal state...* - Even if Greenland was more independent than it is today, I still do not think that it would warrant a demand for complete independence. Maybe, I one day may entertain the possibility of a "federal state", consisting of Denmark, the Faroe Islands and Greenland...but this pushes us far into the next century." (Buchardt, 1992)

Everything points in the direction that the Greenlandic revolution will, for many more years, continue its quiet life within the framework of the Home Rule Act.

(Translated by Eva Jacobsen)

References

Betænkning 837, 1978. *Hjemmestyre i Grønland I-IV.* Copenhagen

Buchardt, Knud, 1992. "Hvor flyder Grønland hen?" *Samvirke,* 3 March 1992

Dahl, Jens, 1985. "New Political Structure and Old Non-Fixed Structural Politics." In *Native Power. The Quest for Autonomy and Nationhood of Indigenous Peoples,* edited by Jens Brøsted et al., Bergen - Oslo - Stavanger - Tromsø: Universitetsforlaget

Dahl, Jens, 1986a: "Greenland: Political Structure of Self-Government." *Arctic Anthropology* 23, 1-2

Dahl, Jens, 1986b: *Arktisk selvstyre - historien bag rammerne for det grønlandske hjemmestyre,* Copenhagen: Akademisk Forlag

Dahl, Jens, 1988: "Self-Government, Land Claims and Imagined Inuit Communities" *Folk* 30

Davies, Jerome D.; Finn Breinholt Larsen; Karen Marie Pagh Nielsen, 1984. *Offentlig styring af olie-gasaktiviteter i Grønland,* Aarhus

Fiskeristrukturudvalget, *Redegørelse fra fiskeristrukturudvalget,* October 11th

Foighel, Isi, 1980. "Home Rule in Greenland." *Meddelelser om Grønland. Man and Society,* No 1 3-18

Gennemførelsesprotokol vedrørende betingelserne for EF-fiskeri i grønlandske farvande fra 1. januar 1990 til 31. december 1994, October 4th

Grønland 1968 ff. *Årsberetning udgivet af Ministeriet for Grønland.* Copenhagen

Hansen, Klaus Georg, 1992. *De kalder os eskimoer. Aspekter af moderne vestgrønlandsk selvforståelse,* Aarhus

Harhoff, Frederik, 1982. "Det grønlandske Hjemmestyres grund og grænser." *Ugeskrift for Retsvæsen* B 101-115

Harhoff, Frederik, 1987. "Dansk rets gyldighed i grønlandske og færøske særanliggender" *Ugeskrift for Retsvæsen B* 347-352

Landstingets forårssamling, 1990. *Forslag til (degrundlag for turismen,* May 2nd, Nuuk

Larsen, Finn Breinholt, 1987. "Scener fra et politisk ægteskab. Den grønlandske regeringskrise august 1986 - juni 1987." *Tidsskriftet Grønland* No 6-7

Larsen, Finn Breinholt, 1987. "Scoresbysund - a hunting community in East Greenland." Paper prepared for 39th Annual Meeting of the International Whaling Commission, Nuuk

Larsen, Finn Breinholt, 1992. "Interpersonal violence among Greenlandic Inuit - Causes and remedies." in *Crime, Law, and Justice in Greenland: Contemporary Perspectives,* edited by Henrik Garlik Jensen. Burnaby, B.C.: Simon Fraser University (forthcomming)

Nalunaerutit - Grønlandsk Lovsamling - Serie A 1976 ff.

Nalunaerutit - Grønlandsk Lovsamling - Serie D 1979 ff.

Royal Greenland, *Årsregnskab 1991,* Nuuk 1992

Råstofforvaltningen for Grønland, 1990. *Rapport fra strategigruppen vedrørende mineralske råstoffer i Grønland,* May 23th, Copenhagen

Sørensen, Axel Kjær, 1983. *Danmark-Grønland i det 20. århundrede - en historisk oversigt.* Copenhagen: Nyt Nordisk Forlag Arnold Busck

Thuesen, Søren, 1988. *Fremad, opad. Kampen for en moderne grønlandsk identitet.* Copenhagen: Rhodos

Culture as Politics:
Experiences from Greenland

Bo Wagner Sørensen

'Culture' has always been a contestable concept in anthropology.[1] It seems
to me, however, that the concept of culture has increasingly been questioned
within the discipline for the last decade.[2] At the same time as 'culture'
seems still more problematic to professionals, it has become a key word
among ordinary people who find it useful and 'telling'. This is certainly a
fact in Greenland. In this paper I want to present a more recent theoretical
perspective on culture, namely culture as politics of everyday life.[3] My
examples are drawn from fieldwork in Nuuk, Greenland, in 1988-89 and
again in 1992.[4] Trying to make sense of my data after returning from the
first fieldtrip, I realized that I could not grasp the often chaotic reality by
use of the conventional idea of culture. This is one in which culture is seen
as an integrated, logically consistent whole held together by a system of
shared meanings that is passed on from one generation to the next. Instead
of finding *the* Greenlandic 'voice', I ended up with many 'voices'
competing with one another to get heard. Instead of finding *the* Greenlandic
perspective, I was confronted with several and not very consistent
perspectives on most subjects and issues.[5] This experience of Greenlandic
culture as fragmented and contested, not just among individuals but even
within the individual, eventually made me focus on the concept of discourse
instead.[6]

This paper is divided in 3 small sections followed by a short summary. I
begin with a sketchy presentation of the most general local notions of
culture and show how these clash with a contemporary anthropological
perspective. Then I proceed to discuss the concept of cultural identity which
is somehow usually seen as a 'natural' offshot of the concept of culture. My
next and last step before the summary is to give a more coherent
presentation and analysis of culture as politics of everyday life.

Local Conceptions of Culture

Most Greenlanders, except for a few well read and reflective persons, think of culture as something one has or possesses. Culture is readily thought of in material terms. It encompasses material objects, artifacts and relics of the past, but it also encompasses Greenlandic language, norms and values, customs, oral tradition, etc., which due to this material thinking become almost reified or quasi-material of kind. Culture, in short, is property.

Culture is also conceived in terms of difference, and to establish a difference it takes a 'cultural other'. The cultural other that everything Greenlandic is compared with and measured against is Denmark for obvious historical reasons. What is believed to be unique for Greenland is presented as part of the Greenlandic culture. Whether differences are real or imagined does not matter. What really matters is what people believe.

The general conception of culture as heritage and something unique is also reflected in the official definition of culture. It seems fair to conclude that culture in a Greenlandic context is mainly used with reference to days of old. The concept of culture has connotations of saving, preserving, storing, nursing, promoting and respecting. Culture is defined as something necessary to have: it is strength, and it provides a life line to the very roots. Culture is a must.

It is no wonder then, that cultural research in Greenland is expected to deal with the so-called cultural heritage. And research is expected to be concerned with documenting and preserving tradition somehow so that future generations can get to know their cultural roots and glorious past and be proud. Cultural research is viewed in a psycho-functional way. It seems that this whole idea of culture as heritage and therefore something to be respected makes it hard to go about cultural research in Greenland in a radically different way.

Like people have ideas as to what counts as culture, they also can tell what does not. I have been professionally interested in issues such as violence, or more specifically wife beating, heavy drinking and sexual promiscuity. These phenomena are without a doubt widespread and part of the contemporary social practice. A common local reaction is that these phenomena can be studied everywhere: abuse is a universal phenomenon and certainly not a specific Greenlandic one. If people apply the concept of culture at all to the phenomena described, they do it to stress abuse as non-cultural, that is as a negation to Greenlandic culture. Abuse is then seen as 'culture out of order'. I, however, have come to the conclusion that wife beating in Greenland may be seen as the 'natural' outcome of a specific cultural logic and social practice.[7] The logic, though, is not one that can be found empirically. The cultural logic is an analytic abstraction based on interpretation of data.

What I find interesting about the general Greenlandic conception of culture is mainly its political implications. If we stick to wife beating as an example, most local people seem to try to think and explain away the problem. When people do not take the problem seriously by making believe that it is not really a *Greenlandic* problem though it actually happens in Greenland, I think they are about to sacrifice part of the Greenlandic population, namely the victims of violence. And the sacrifice in fact takes place in the midst of ethnopolitical rhetorics of solidarity. People are prone to think away violence because it does not fit with the Greenlandic high culture.

My next step will be to consider another, but parallel, aspect of culture as politics, namely how people 'talk culture' in everyday life. This culture talk often takes the guise of a preoccupation with cultural identities.

Cultural Identity

Cultural or ethnic identity is a concept which, like culture, often has quasi-property connotations: an ethnic identity is something you have, and if you have not, you are bad off and better try to get one. Identities are perceived in terms of wholeness. The ideal is the unquestioned and solid identity, and people are believed to be in severe trouble if they are between cultural identities. Often they are indeed, but that is not because of individual psychological needs. Rather their problems are due to social expectations. I would say that ethnic identity is not something you naturally have or need to have. Ethnic identity like other identities are primarily socially imposed, only secondarily individually chosen among identities available. Another thing about ethnic identity is that it is not present all the time. Rather it is only present when it is evoked.

One Friday afternoon in Nuuk, I was on my way home from work. Walking through the main street I happened to be spotted out by a Greenlandic woman in her mid-twenties whom I did not know. She somehow did not like me or rather the Dane she saw in me, so when I passed by her she cried in a shrill voice: "Why are you Danish?" Her rather inarticulate but anyway clear message was repeated twice in a still more powerful voice as I walked away. That Friday, until I was identified as a Dane and thus as someone not really belonging, I had not given 'my' cultural identity a single thought all day. I had been *me* at work and later on *me* on my way from work along with many other people.

The fact that my identity was being evoked that Friday and has been now and then in my politics of everyday life, is quite parallel to what other people in Nuuk experience whether they be Greenlanders or Danes. But usually it happens in a subtle way rather than the dramatic way just described. In Nuuk, people distinguish between Greenlanders and Danes, but the category of Greenlanders is certainly also differentiated. Greenlanders living in Nuuk belong more or less. Some, the Nuummiut, are of the town, others just live there. Some are more 'real' Greenlanders than others. Some are very conscious of their cultural knowledge and perhaps especially their linguistic skills while others just use their language without giving it much thought. Some are of 'pure blood', whereas others are of 'mixed blood'. The Danes are also differentiated. Some are locals and thus belonging somehow, others are known as temporary inhabitants.

Now, categories are not simple reflections of real differences. Whether a Greenlander is more or less 'real' Greenlander depends for instance on definition and context. Real Greenlandicness calls for evaluation. The interesting thing about categories is therefore how they are used, that is in what circumstances and by whom. It is to categories in use that I now turn.

Belonging More or Less

It seems to me that people in Nuuk mark off real Greenlanders from less real Greenlanders and finally Danes. Precisely what makes a real and a less real Greenlander is not possible to say. The point is that these categorizations are negotiable. It is also a fact that the different identities springing from the categorizations are accompanied by certain rights and claims. That is why most people in Nuuk seem to be taken up with identities. The Greenlandic quest of identity has to a high degree been explained in literature as a deeply felt and pressing psychological need, a need to find and stabilize the cultural roots. However, I believe that the psychological explanation has to do with our inclination to turn to a psycho-functional frame of reference instead of a sociological one.

The preoccupation with cultural identities and whether they are solidly grounded or not has to do with the fact that cultural identities are officially recognized which means that certain rights and privileges are accorded people on the basis of these identities. After Home Rule was introduced the official policy has been to replace what is seen as non-local workers and employees by locals. In actual life it means to a high degree to replace Danes by Greenlanders. The issue of identity often seems to be actualized in situations of perceived shortage. Identity has to do with boundary-making, and in situations of perceived shortage identities are evoked to establish a difference between people who belong and people who do not 'naturally' belong. The first category can claim certain 'natural' rights and privileges whereas the second category can just hope to get a share, but cannot claim it as a birthright. It seems that rights and privileges are intimately connected with belonging, and I think that this fact also explains that Greenlanders differentiate between real and less real Greenlanders. This differentiation is another way to put the question of belonging which ultimately is a way of communicating social status.

Language is a very concrete marker of difference between Greenlanders and Danes. That is why linguistic competence, along with more diffuse and questionable cultural knowledge such as knowing *the* Greenlandic way of thinking, is given a central position when Greenlanders argue that jobs in Greenland should be reserved for Greenlanders, in the long run at least. Greenlandic in itself becomes a symbol of local competence, of being capable of functioning at the place of work and elsewhere, whereas other qualities are usually toned down. Greenlanders' claims to perceived limited local goods are made through assertions of linguistic competence which operates as an expression of belonging and well-functioning.

Greenlandic, however, also operates as an expression of real Greenlandicness. Among Greenlanders themselves, linguistic competence is used in an assertive way to legitimate one's social position or to make claims. People who do not speak the language are supposed to have certain serious shortcomings. Purists among Greenlanders do not hesitate to correct and scold others who do not speak fluently or correctly, whatever that is. Moreover, they emphasize that the very Greenlandicness is somehow embedded in the language. It follows that if you do not know the language in depth, you are not one of the real Greenlanders.

But language is not the only standard. Once again, purists among Greenlanders try to rule out careless and abusive elements of the local population on grounds of their non-cultural or uncivilized way of living. However, the purist or cultural elitist critique of others is no guarantee that they themselves lead a sober life. The abusive elements on the other hand may criticize the cultural elite for being too Danish-like, materialistic, stingy and conceited. The cultural elite to a certain degree try to forestall and avert such critique by behaving heartily and paternalistic. Their interest in cultural matters are moreover usually presented in self-sacrificing terms as all-important work done for the benefit of the Greenlandic people.

Although possibly all Greenlanders of today are of 'mixed blood', the direct children of a Greenlandic-Danish union often experience that they are seen as neither/nor and therefore bound to have problems. In a social context where cultural identities are conceptually clear-cut, it is no wonder that the mixed-bloods are often expected to be persons waging war on themselves as their two 'sides' are seen as fundamentally non-compatible. However, I also met a few of these in-between persons who did try to react against the oppression that is implicit in the expectations based on the categorical thinking described. One of them, a woman in her late twenties, consciously did not want to choose side in order to pass as Greenlander. When Greenlanders asked her about her ethnic identity, she deliberately would say: "I am a bastard." Referring to Greenlanders and Danes in conversation, she would also count herself out of both categories. It seemed to me that she refused and fought against categories in use which she herself had experienced as oppressing and degrading. She also deliberately fought the cultural imperative that Greenlanders have to speak Greenlandic. She spoke some when she had to, stressing the aspect of language for communicative use, but when she felt that she was expected or provoked by some Greenlanders to speak Greenlandic, she turned to Danish. Language had become part of the politics of everyday life.

Few people seem to be as courageous as the woman just mentioned. Another woman about the same age with all-Greenlandic parents but a Danish-like adolescence speaks some Greenlandic while she speaks Danish fluently and on a much more sophisticated level. She was at a party held by a Greenlandic politician and most of the guests belonged to the leftish cultural elite in Nuuk. Being a non-properly Greenlandic speaker and though very Greenlandic looking according to her own definition, she dared not speak all night as she was afraid to give away her linguistic incompetence as regards Greenlandic. The way I see it, her very fears of the other guests' reactions are quite 'telling'.

Culture as Everyday Politics:
Whose Culture Is It Anyway?

Using data from Nuuk, I have tried to present another way, and in my opinion more promising way than the conventional one, of dealing with 'culture'. Culture should be viewed from a social actor perspective as such a perspective can show how culture in its different manifestations is lived and made use of. The way people think of culture, and the cultural categories they make use of, structure their social universe and have a very direct influence on everyday life. Culture becomes politics as culture is used to manipulate, monopolize, negotiate and make claims in a world of social inequality. Who really belongs? Whose culture is it anyway? Cultural identity and competence of one kind or another takes on such significance and centrality precisely because it is inextricably bound up with social status and inequality.

The analytic concepts applied in social science research in Greenland are all too often used unreflectively, and the theoretical perspectives are often somewhat outdated. That is why a reified notion of culture is still common; a notion which many researchers and most locals hold in common.

Notes

1. See, for example, Keesing 1974.
2. See, for example, Barth 1989; Marcus & Fischer 1986. The Danish anthropologist Kirsten Hastrup also has written extensively on the concept of culture. See, for example, a recent article by Hastrup (1991) for further references.
3. I have been inspired by other anthropologists working in this tradition, especially Abu-Lughod & Lutz 1990, Collier 1988, and Strathern 1982. See also the Danish anthropologist Anne Knudsen (1989).
4. I have worked and lived in Nuuk since February 1992, and experiences and notes from this period of course also affect my presentation.
5. See Sørensen 1990, 1991.
6. See Sørensen 1992.
7. See Sørensen 1990, 1992.

References

Abu-Lughod, Lila & Catherine A. Lutz, 1990, *Introduction: Emotion, Discourse, and the Politics of Everyday Life*. In C.A. Lutz & L. Abu-Lughod (eds.): Language and the Politics of Emotion. Cambridge: Cambridge University Press.

Barth, Fredrik, 1989, *The Analysis of Culture in Complex Societies*. Ethnos 54(3-4).

Collier, Jane Fishburne, 1988, *Marriage and Inequality in Classless Societies*. Stanford, CA: Stanford University Press.

Hastrup, Kirsten, 1991, Antropologiske studier af egen kultur. *Norsk Antropologisk Tidsskrift 1*.

Keesing, Roger M., 1974, Theories of Culture. *Annual Review of Anthropology Vol. 3*.

Knudsen, Anne, 1989, *Feltarbejde blandt levende og døde: Korsika - litterært fænomen, etnografisk objekt, ferieø og samfund*. I Kirsten Hastrup & Kirsten Ramløv (red.): Kulturanalyse: Fortolkningens forløb i antropologien. København: Akademisk Forlag.

Marcus, George E. & Michael M.J. Fischer, 1986, *Anthropology as Cultural Critique*. Chicago: The University of Chicago Press.

Strathern, Marilyn, 1982, *The Village as an Idea: Constructs of Village-ness in Elmdon, Essex*. In Anthony P. Cohen (ed.): Belonging: Identity and Social Organisation in British Rural Cultures. Manchester: Manchester University Press.

Sørensen, Bo Wagner, 1990, Folk Models of Wife-Beating in Nuuk, Greenland. *Folk 32*.

Sørensen, Bo Wagner, 1991, Sigende tavshed: Køn og etnicitet i Nuuk, Grønland. Tidsskriftet *Antropologi 24*.

Sørensen, Bo Wagner, 1992, *Magt eller afmagt? Køn, følelser og vold i Grønland*. Ph.D.-afhandling, Institut for Antropologi, Københavns Universitet.

Potential for more Diversified Trade and Industry in Greenland

Jørgen Taagholt

For millennia Eskimo tribes have lived out an isolated existence in harmony with their harsh natural surroundings. The scanty population in the vast territories where they lived did not disturb the ecological balance with its hunting and fishing. This is what we today call "sustainable development" - the theme of this conference.

Like the animals they hunted, the people had to adapt to natural change, or had to emigrate to follow the animals, and many succumbed in the struggle to sustain life in marginal subsistence conditions. History shows us that the living resources in a geographical region like Greenland can sustain a maximum population between ten and fifteen thousand people.

Hans Egede's work in Greenland began a slow process of European cultural influence where the Christian culture was unable to accept the ancient Inuit social structure which required that when old people, orphans or the sick became too much of a burden to the community, it was the duty of the individual to choose death to ensure the survival of the group. The interests of the society or community took priority over those of the individual.

The Brundtland Report "Our Common Future" defines sustainable development as development that meets the needs of those now living without compromising the potential of subsequent generations to meet their needs. In terms of this definition, the Inuit's thousands of years of development in harmony with nature, based on the living resources of nature, cannot be called "sustainable", since as we have seen the living resources have been unable to meet the needs of a growing population.

Post-war development

Until World War II, Greenland was on the whole self-sufficient, with an economy based on hunting and fishing supplemented by mining at the cryolite quarry in Ivigtut and a limited amount of sheep-farming. During the Second World War Greenland suddenly acquired a strategic position as a result of technological developments. This led to the establishment of several air bases, defence installations for the protection of the cryolite quarry at Ivigtut, and a number of meteorological stations. After the war there was a growing local desire in Greenland to open up the country after the isolationist policy of the colonial era, and to allow Greenland to benefit from the technological developments that the world war had given the Greenlanders the opportunity to see at close quarters - as spectators (Taagholt 1980).

The first five HF radio stations in Greenland were established in 1924-26, but several stations were set up during the war by the Danish/Greenlandic authorities in Greenland with the financial support of the USA, because of the need for meteorological reports (Bach & Taagholt 1976, p. 205). A period of favourable climatological conditions led to high fish stocks in Greenlandic waters, and as part of the development programme of the sixties, the traditional hunting and fishing community in the most densely populated area, South West Greenland, was gradually transformed into a modern fishing community where the kayak was replaced by the trawler (with the fishing cutter as the transitional stage), and modern fish processing factories were established to produce high-quality deep-frozen fish products for the international market.

Industrialized fisheries

Industrialized fisheries and the fishing industry became the staple economic activity in the post-war years, as referred to in the report of the Greenland Committee of 1960 (normally abbreviated G60). The outcome of this development has been a total transformation of the structure of Greenlandic trade and industry, and fisheries are, and will remain for many years to come, the foundation of the Greenland economy. However, at present Greenland is experiencing how vulnerable society is when it is wholly dependent on a living resource like fish. The one-sidedness of the trade and industry structure raises a number of problems - not least in a situation where the fishing industry is dependent on a set of factors over which the Greenlandic community has little or no control, such as international trade agreements and quota systems, and climatic variations which could totally change the basis of commercial fishing. Cod fisheries, for example, were reduced by about 80% between 1989 and 1991. The reason for this is that the 1984-85 year stocks which formed the basis of fisheries have left Greenlandic waters *(Nyhedsbrev* No. 7, 1992; Lyck & Taagholt 1987).

In other words, the living resources cannot meet the ever-growing needs of the population, so if they are to aim at sustainable development, it is necessary to look at the potential for kinds of trade and industry which are not based on the limited living resources of the region. In the following an attempt will be made to assess the potential for alternative trade and industry in the light of technological and geopolitical developments.

Radio stations and weather reports

Building and public works, increased industrial activity and the development of the infrastructure also meant a need to introduce modern telecommunications. The very high operating costs of manned radio and weather stations in desolate areas of Greenland meant that Denmark was unable to maintain its comprehensive manned service, and this led to the closure of a number of North East Greenland stations in the 70s.

Since Denmark had a commitment with the UN World Meteorological Organization to provide meteorological data to operational and scientific users, the Commission for Scientific Research in Greenland took the initiative in 1972 to start up a project in which various Danish research institutions cooperated in developing automatic meteorological stations, so-called UGO (Unmanned Geophysical Observatory) stations (Taagholt 1977).

As a result of this project, in the period 1972-85, over twenty stations were set up, and in 1977, in cooperation with NASA, the first satellite link from Greenland was established when the UGO at Station Nord was furnished with equipment for transmitting data via the satellite NIMBUS 6, which orbited around the polar region at an altitude of 1100 km. On the basis of the positive results from this, other UGOs were equipped with satellite transmission equipment, operating with the satellite communications system ARGOS, which works as a RAMS (Random Access Measuring System), where a data burst is re-transmitted every minute, for example, and data is received and relayed from the satellite to the ground station which was established at Søndrestrømfjord/Kangerlussuaq. From 1979 on, the UGO stations south of 70°N were further equipped with transmission equipment.

Telecommunications

Concurrently with this project, the Greenland Telecommunications Authority worked from 1975 on the establishment of a modern analog radio linkage in Greenland, which would ensure reliable telecommunications among the West Greenlandic towns and settlements (Bach & Taagholt 1976, p.199). Possibilities studied included the use of knife-edge diffraction, which allows for longer stretches and the placing of relay stations in valley bottoms, but this solution was only used between Ammassalik and Kungmiut in East Greenland, while the West Greenland analog chain is based on "line of sight", with unmanned relay stations placed on the West Greenlandic mountaintops. Since 1978 the Greenland Telecommunications Authority has used a geostationary communication satellite of the INTELSAT type for telecommunications between Denmark and Greenland, and since 1982 geostationary satellite communication has been used to link the West Greenland UHF chain with isolated districts like Thule and Scoresbysund. Remote, isolated communities which formerly only had contact with the outside world a few times a year now have telephone, fax and regular air links (Taagholt 1983 - A).

Operational use of satellites

Because of the huge geographical extent of Greenland, satellites are used today, not only for routine links between the West Greenland UHF radio chain, the outlying districts and Denmark, but also for remote sensing, which has become an important tool in a number of scientific and operational disciplines. At the initiative of the Commission for Scientific Research in Greenland a development project was started as early as 1967 at the Technical University of Denmark, dealing with radar technology and digital image processing. The best known application today is digital satellite pictures used in meteorology, but in future they will increasingly by used in hydrology (hydroelectric power) and oceanography (climatic surveys), geology (mineral exploration), and for example for identifying the grazing potential for sheep and reindeer in Greenland. In connection with the projects in Greenland, the Technical University has developed new airborne radar systems which are ready for trials. These allow radar images to be registered from a 250 km^2 area at a resolution of 2 m in just 100 seconds. It has subsequently taken hours to process such amounts of data in computer centres, but recently-developed Danish equipment now makes concurrent display of the pictures in the aircraft possible - this is probably the fastest equipment in the world (Taagholt 1993).

Advanced telecommunications

The rapid development of Greenland brings with it a natural need for an expansion of telecommunications. When the internal telephone networks of the Greenland towns were linked up by the UHF radio chain, telephone and telex connections came into daily use in business, administration and private homes. Greenland is probably the first region in the world where fax has become part of everyday life for private communication among family members. At the same time a need has arisen for the transmission of large amounts of data on, for example, meteorological and oceanographical conditions, and digital satellite images for use in operational services like patrolling the Greenland waters, fisheries inspection and the ice service. The constantly growing demands on the capacity and quality of the UHF chain have led to a need to implement a digital radio chain to supersede the analog one.

I 1991 the Greenland Home Rule Government signed contracts for supplies for the new radio chain, according to which Siemens Denmark is to supply radio equipment for the 8 GHz band, the firm ANDREW is to supply the aerials, and the Jutland telephone company is to supply the monitoring equipment. The project is divided into five phases. The first phase, covering the stretch from Nuuk to Paamiut, will be concluded in 1992. In the next few years, the project will gradually work southward as well as northward, and it is expected that the whole stretch from Frederiksdal to Uumanaq will be completed by 1996. The radio chain will then have reached a level that will also satisfy the users of tomorrow.

Commercial potential of data processing industry

Electronic data processing is increasingly being used in more and more areas. With Greenland's modern telecommunications systems, for example, banks in Greenland can be linked directly with corresponding computer equipment in Denmark, which makes financial transaction both quicker and safer. In many areas large amounts of data are processed today in computer systems - for example in public administration, energy supply, trade etc. In construction and public works, calculations, drawings and specifications etc. can be transferred over linked computer systems. However, large volumes of historical data from jut a few years ago - for example economic statistics, information on fishing catches, meteorology etc. - is only available in tabular form and is not machine-readable. Experience has shown that Greenlandic telegraph operators were very fast and conscientious in their work with the morse key, work that has now been totally replaced by the modern telex or fax machine. Given this good experience, it should be possible in Greenland, with just a little investment, to build up firms with the right specialized environment who could undertake to enter data from various specialized areas in computers. In relation to the study of climatic development, for example, there is a need to enter 100 years of meteorological data. With the present infrastructure in Greenland, such firms could,

222

against small investments, help to increase employment, at least for a few years.

Mineral resources

Our knowledge of mineral deposits in Greenland can be traced back to Hans Egede's time, about 1840, but production has been modest.

At the end of the nineteenth century about 7,000 tons of graphite were quarried at several sites, and copper was mined in several places - for instance at the Frederik VI copper mine near Julianehåb/Qaqortoq. In the period 1936-40, and again in the 1960s, marble was quarried at Maarmorilik and used to face several buildings (e.g. the National Bank) in Denmark. More important was the cryolite production at Ivigtut, where about 3.5 million tons of cryolite were shipped out in the period 1856-1980. On Disko Island in 1924-72, about 600,000 tons of coal were shipped out, and in 1952-55 about 560,000 tons of lead and zinc ore were mined at Mestersvig in East Greenland (Bach & Taagholt 1976, p. 69).

The mining activity that has been most important to technological and social development in Greenland in the post-war years is undoubtedly the lead and zinc mining at Maarmorilik.

The Maarmorilik mining complex is an example of a successful industrial activity. In 1973-90 2.9 million tons of zinc and lead concentrate were produced. The mining made a contribution to the Danish/Greenlandic economy of about DKr 500 million in profit tax and concession charges. The mining company Greenex A/S was able to involve local manpower to a considerable extent - up to 45% of the workforce was Greenlanders. It was the first company in Greenland to introduce equal wages for the same work, irrespective of whether an employee was from outside Greenland or was a local inhabitant. The company was very important for local education and training in various branches of mining and had a great influence on the building up of the infrastructure in the local area around Umanak. After the lead and zinc deposits had been exhausted the mining activity was discontinued in 1990. Recently-discovered lead and zinc deposits north of

Maarmorilik mean that the potential for new mining activity in the area is being assessed at present.

The Greenland Geological Survey, today under the Ministry of Energy, is responsible for overall geological surveying in Greenland, and this provides the essential background for more detailed surveys by prospecting firms of the quality, extent and quarrying/mining feasibility of identified deposits.

We are today well aware that there are deposits of valuable minerals in Greenland, such as uranium (at Narsaq alone deposits have been discovered which could theoretically supply enough uranium for perhaps fifty years of Danish power consumption); gold (deposits in East Greenland, because of low quality, have no commercial interest today, but newly-discovered deposits in South West Greenland, of much higher quality, are now being studied in more detail); platinum (which, with palladium, has been found along with gold deposits in East Greenland); molybdenum (the deposits in East Greenland are in a very inaccessible area); chromium (the limited deposits in South West Greenland are nevertheless the most extensive occurrence within the NATO area); iron (large deposits at Godthåbsfjord have no commercial interest in their own right, but a putative steelworks in West Greenland, exploiting the area's immense hydropower potential, combined with Greenlandic deposits of alloy materials like chromium, molybdenum, wolfram and others could form a basis for advanced steel production at some time in the future); silver; lead; zinc; and rare earths such as niobium (which could perhaps be used in the production of superconductors). In addition, big international oil companies, in collaboration with the Danish/Greenlandic company Nunaoil, are carrying out seismic surveys of the Continental Shelf in both West and East Greenland to identify geographical areas with favourable conditions as potential sources of oil and natural gas.

Hydroelectric power

Less than 20% of the hydropower reserves of the planet are exploited at present, primarily because the potential resources are situated far from the users. Pilot studies in Greenland have indicated the potential for building a number of hydropower plants which, at a conservative estimate, could produce 12-15,000 GWh a year by exploiting the run-off from meltwater lakes in fell terrain at the edge of the ice sheet (GTO 1975). This volume of production corresponds to about 50% of Denmark's present consumption. More than 80% of Greenland is over 2,000 m above sea level. If the water could be channelled from the natural meltwater lakes on the ice sheet out to the mountain lakes on the periphery as a reservoir for hydropower production, the potential would be multiplied many times (Taagholt 1989).

The potential resources in Greenland amount to about 1/1000 of potential global hydropower resources, so results from Greenland would be of great international interest.

Global environmental concern may mean that in the near future the industrialized world will introduce an environmental tax on energy consumption, based on the use of fossil fuels (coal, gas, oil). Such a tax would suddenly make remote hydroelectric resources economically and environmentally attractive.

Buksefjord hydroelectric power station

The first major hydroelectric power plant in Greenland, at Buksefjord, south east of Godthåb/Nuuk, will be operationalized in 1993.

The Greenland Energy Authority of the Greenland Home Rule Government is the client for this plant, Greenland's biggest-ever public works venture. After international tenders were invited, a general contract for the construction and the first fifteen years of operation was awarded to Nuuk Kraft ANS, a consortium formed by Norwegian firms (Kværner Brug A/S, EB Energi and EB Kraftengineering) in collaboration with the Danish/-Greenlandic firm Atcon A/S. The contractor undertakes to spend 25% of the construction price on subcontracting and purchases in Greenland (R. Andersen and M. Høgsted 1992).

The reservoir is the roughly 70 km² mountain lake Kangerdluarssungup taserssua, whose natural surface is about 250 m above sea level; a 150 m long, 15 m high dam produces a maximum head of 261 m. A 56.5 km high-voltage cable including the world's longest single span - 5,376 m - over the Ameralik fjord, is being laid by Bentonmast A/S, one of Norway's most experienced contracting firms in transmission cables, with special experience of long fjord spans. This will mean an energy surplus for the Nuuk area.

The reservoir, with an area of about 1.9 b. cubic metres, has a high reserve capacity, and can contain enough water for more than five years' power production at the present installed output from two 15 MW Francis turbines and two 18 MVA generators, corresponding to an annual production of 185 GWh. Of this, 55 GWh will be reserved for lighting and power in the Nuuk area. Most of the remainder will be used for production of district heating using electric boilers.

In financing terms, the power station at Buksefjord represents an innovation. While public works have hitherto been financed quite traditionally (by the Danish and Greenlandic public sectors, until 1985 often with subsidies from the EC infrastructure fund), the hydropower station is financed by the Greenland Home Rule Government (DKr 200 million), a loan from the Nordic Investment Bank (DKr 400 million) and international loans guaranteed by Norsk Eksport-finans (DKr 450 million).

Industrial use of hydroelectric power

At the moment, as the first industrial customer for hydroelectric power, a plastic goods factory is being erected in Nuuk, and the remainder of the power produced will be used, until other industrial customers appear, as backup electric heating in the relatively new oil-fired district heating plant in Nuuk. It will however be interesting to have an assessment of how the energy from the Buksefjord plant could contribute to the power supply in other urban areas in Greenland. Because of the great distances and the relatively small amounts of energy required, transmission as electric power in high-voltage cables is not feasible. The Greenland Home Rule Government has therefore, through the Greenland Energy Authority, initiated a project to study the possibility of energy transport in the form of energy-bearing material - for example liquid hydrogen, ammoniac or some more suitable material (Taagholt 1983 - B).

Much of the lead and zinc concentrate from Maarmorilik was shipped from Greenland to Belgium or Finland. At present large amounts of concentrate are still being shipped from Canadian mines for processing in Europe. Platinova A/S in Nuuk is now working to study the possibility of locating a zinc refinery at Nuuk. This would mean that freight costs from Canada to Europe would be reduced, electricity from the Buksefjord power plant could be exploited, and employment and the economy in Greenland would be generally improved.

From the Canadian Arctic zinc mines Polaris, Nanisivik and as of 1996 also from Izok Lake, about 600,000 tons of zinc concentrate will be shipped per year from Arctic Canada via the Davis Straits to the refinery in Europe, with the resulting high transport costs.

Platinova A/S is now carrying out preliminary investigations with Canadian mining companies. It is thought that a zinc refinery at Godthåbsfjord with a production capacity of 100,000 tons of pure zinc a year will require an investment of about DKr 1.5 b. According to the present plans the power requirement could amount to about 400 GWh a year, which would mean an extension of the power station beyond the present extension plans. If the technical and economic assessments show that production can be made profitable, it is believed that the necessary capital can be raised. Besides the major initial construction work and the employment that will result from it, the operation of the plant will provide employment for an estimated 350 people.

The surveys done so far seem to show that the need for power will even exceed the extra capacity of 20 MW can provide. Studies have therefore been initiated of the potential for extra water supply to the reservoir from a nearby lake higher up; if this can be used, there will be enough water for a total 70 MW installed output, corresponding to an annual production volume of about 475-525 GWh. The present tunnels to the turbine pipes, the actual power station building and the transmission cable, can accommodate such an extension, but an extra two 20 MW turbines, generators and a tunnel to the higher lake will cost an estimated DKr 400 million.

Development of society

In a single generation - about thirty years - Greenlandic society has thus changed from an *ecologically* very highly developed, specialized hunting society into a *technologically* advanced industrial society using the technology of today - and indeed of the future. As a curiosity I can mention that the first public TV transmission in Denmark took place in Narsarsuaq, and that Denmark's tallest building is the 400 m high aerial tower at Thule AB. International military and industrial interests have left their mark on a development which in many ways has benefited Greenlandic society. Better living conditions have helped to eliminate tuberculosis, and average life expectancy has increased substantially.

In recent years external factors have been crucially important for Greenland. Lower sea temperatures mean far fewer fish, century-old mining activities have ceased after the known deposits were exhausted, and geopolitical factors have meant a decline in military interest, with the resulting cessation of the contribution of the military to infrastructure development and maintenance. Greenland is therefore in a very difficult situation economically and in terms of employment (Morits Andersen & Taagholt 1993).

Tourism

Greenland has a unique asset in its thrilling and dramatic natural landscape.

Many people have an erroneous picture of Greenland as a snowy, icy waste. The real picture is one of flowering plains with fertile grazing for sheep, reindeer and musk-oxen, and Alpine mountain terrain where towering glaciers spit icebergs into the fjords. The Greenland Home Rule Government is convinced that we are facing an international shift in tourist interests. Tourists are growing increasingly tired of overcrowded roads, cities and bathing beaches and of the increasing pollution. Today many of them would prefer the challenges of a mountain hike, the close encounter with a desolate, breathtaking natural landscape, its flowers and pure, clean air, and a visibility of more than 100 km. Mass tourism would be too much of a burden for the Greenlandic outlands, but when the North Atlantic flight route network in future does not stop at Greenland, but is linked effectively with Canada and the USA, "stopover" tourism of the type amply exploited in Iceland at present will become a possibility; passengers on route between North America and Europe, at no extra expense, will be able to spend a few days in Greenland. In 1992, to promote tourism in Greenland, the Greenland Home Rule Government established "Greenland Tourism", a firm wholly-owned by the Home Rule Government with the aim of increasing the number of tourists in Greenland. By the year 2005, for example, it should be possible to reach a volume of 35,000 tourists a year.

Future prospects

It is well known that Greenland has deposits of valuable minerals such as uranium, gold, silver, platinum, molybdenum, chromium, lead, zinc and rare earths like niobium. The problem, however, is finding the right, commercially viable deposits. Because of the high level of costs in Greenland, a deposit should preferably include rare or precious minerals, be very large, and be of a high quality compared with what can be produced by similar activities elsewhere in the world. In 1991, to promote interest in mining, the Greenland Home Rule Government adopted a new mining act, which safeguards the rights of prospecting firms better, and lays down conditions, terms and requirements in connection with raw material exploitation.

Greenland society increasingly wants to be involved in prospecting and exploration work and the related service activities. Large potential hydropower reserves form a background for assessing the technical and economic viability of the production of aluminium, artificial fertilizers, energy-bearing materials, smelting works for metal processing, or other exploitation of Greenland's raw materials and renewable energy sources, using the technology of tomorrow to avoid unintentional harmful environmental effects.

The implementation of projects like these requires big investments. Although in 1985 Greenland opted out of the EC, overall geopolitical interests may encourage the exploitation of renewable Arctic energy sources for global environmental reasons, and may thus help to promote necessary industrial development in Greenland.

References

Andersen, Raae & Høgsted, Mogens, 1992, Vandkraftværk i Nuuk, *Byggeindustrien nr. 7*, København.

Andersen, Anne Morits & Taagholt, 1993, Geopolitical developments and their significance for Greenland and the circumpolar Arctic area, *Proceedings from Nordic Arctic Research Forum Symposium*, Copenhagen.

Bach & Taagholt, 1976, *Udviklingstendenser for Grønland*, Nyt Nordisk Forlag, København.

GTO, 1975, *Lokalisering af vandkraftressourcer på Grønlands vestkyst*, København.

Lyck & Taagholt, March 1987, Greenland today - economy and resources, *Arctic*, Calgary, Vol. 40, No. 1, p. 50-59.

Nyhedsbrev nr. 7, 1992, Økonomidirektoratet, Grønlands Hjemmestyre.

Taagholt, 1977, Problemer omkring de grønlandske vejrstationer, *Grønland nr. 6*.

Taagholt, 1980, Greenland and the future, *Environmental conservation, Vol 7*, No. 4_, p. 295-299.

Taagholt, 1983-A, Geofysiske forholds betydning for radiokommunikation og radionavigation, *Naturens Verden 9*, p. 193-216.

Taagholt, 1983-B, Grønlands ressourcer og muligheder for nye udnyttelser, *BP Energiårbog, Futuriblerne*, p. 121-165.

Taagholt, 1989, Future industrial development in the Arctic and growing global environmental concern, *Mining in the Arctic*, Bandopadhyay & Skudrzyk (eds), Balkema, Rotterdam, ISBN 90 6191 899 5. p. 33-42.

Taagholt, 1993, Avanceret radarteknologi og digital billedbehandling. *Naturens Verden*, p. 64-72, Februar 1993.

The Conditions for Employment of Greenlanders in the Home Rule Administration - Especially at High Official Level

Mette Marie Skau

The background to the following is a report submitted in connection with my final degree from Aalborg University. It was written from 1988 to 1989, and some of the conclusions I made then may now seem less important. Nevertheless, I think that the problems I described then still exist in many ways.

It is wellknown that Greenland has a lack of well-educated labour. As regards the Home Rule Administration, you will see that most positions at high levels are filled with Danish employees. And you can see that since 1979, where the Home Rule was introduced, the number of Danish employees has increased every year, although as compared to the number of Greenlandic employees, it has been smaller over the years.

An increase of Danish labour has especially been seen in connection with transferring political authority and administrative tasks from the Danish state to the Home Rule. In the period 1984 to 1987, where PROEKS (fishing/ production), KNI (trade) and GTO (housing, etc.) were transferred, the number of Danish employees increased from 318 to 1035. In other words, there has been a disproportion between the Greenlandic labour abilities and the demands of the Home Rule Administration.

Politically this development or trend has been considered as a problem. Especially when considering what the political movements for independence in the years before 1979 were demanding. What they wanted was a Greenland on Greenlandic conditions, more specific included oppositions against the amount of leading administrators from Denmark. The Greenlandic politicians lack influence. And there was an opposition against the neglecting of the Greenlandic language in the schools. Having employed all these Danes did not suit the political work, or image, and therefore decisions have been taken to work against this development. As examples of these activities can be mentioned internal courses for the staff in the Home Rule Administration, the establishment of an administrative degree at the university in Nuuk, and laws concerning regulations of the recruitment of (Danish) labour.

In the following I shall present very shortly some of the theoretical frames or angles for discussing the conditions for Greenlanders to be placed in all positions in the Home Rule Administration. These were at the beginning mainly inspired by Jytte D. Mikkelsen ("undersøgelse af de grønlandske erhversforhold - med udgangspunkt i afhængighedsforholdet mellem Grønland og Danmark", AUC 1985) who compared Greenland with post-colonial societies in the Third World, and Jens Dahl (" Arktisk selvstyre - historien bag og rammerne for det grønlandske hjemmestyre." 1986) who saw the Greenlandic Home Rule Administration as inherited from Denmark and overdeveloped.

As regards post-colonial societies let us for a start have a look at their characteristics. Gorm Rye Olsen (" Perifere samfund - det er mange ting." Grus 1980) emphasizes four signs:

(1) They have very complex economic structures. Besides the capitalistic structure, there are also non-capitalistic elements.

(2) The above-mentioned indicates that often more complex class-structures exist than in the fully developed capitalist countries.

(3) The presence of a foreign bourgeoisie with strong economic positions.

(4) The presence of overdeveloped state institutions.

The last-mentioned is explained by the interests and great influence in the post-colonial state of fully developed countries and on the other hand the interests in an intravert development the post-colonial states.

The post-colonial state is here only seen as a product of the economic conditions and the class structure. Another angle is the one of Hamza Alawi (" Staten i perifere og postkoloniale samfund: Indien og Pakistan." John Martinussen 1980). He says that there is a direct connection to the former authorities in the post-colonial state. This means that there will be a tendency to continue using administrative structures and traditions from the period before independency.

What I just described helps especially to pointi out the theoretical frames for the economic and political reasons for the Home Rule in Greenland. But it does not in a sufficient way explain the conditions inside the administration.

Seeing the Home Rule Administration as an inherited ministerial structure from Denmark a theoretical background must also contain the signs of a bureaucratic administration. Referring to Max Weber the most important of these would be the following:

-        The formal principle about certain authority connected to certain tasks.

-        The hierarchic structure of the administration.

-        Allocation of resources to smaller units is calculated very closely.

-        A specific education is expected.

-        Superior rules for discarge of offices.

-        Formalized career-making.

What is not evident from the above are the informal mechanisms. Hartmut Haussermann ("Bureaukratiets politik. Indføring i den statslige forvaltningssociologi." 1977) describes these as structural conservatism. So saying he means that the units in an administration have a tendency to accentuate their own importance to maintain their share of the resources at the expense of a certain political decision. And that is mainly being done by referring to their specific experience and knowledge.

Besides the formal and informal signs of a bureaucratic administration one more angle is needed. In spite of the similarities of the Home Rule Administration with traditional bureaucratic organizations it is still special regarding its political and economic background.

This angle would be the one of Søren Christensen and Jan Molin ("Organisationskulturer, kultur og myter." 1983). They have worked with the idea of an organizational culture. This is explained (briefly) as some certain patterns of myths, norms and routines that get obvious through 'generations'.

Looking at the labour in the Home Rule Administration as a whole these 'obvious patterns' might not be considered that obvious by the greenlandic labour. Therefore a tendency to conflict in relation to the tranferred structures in the administration might display. Assuming that there is not a quite suitable organizational culture you thus also have a frame indicating that a more advantageous change for the Greenlandic labour might take place in the administrative structure.

In the following I shall give a very short presentation of my conclusions.
As I see it, the Greenlandic Home Rule 'State' can be described as being overdeveloped. It is not fundamentally a product of the Greenlandic society; neither does the ecomic development base exclusively on Greenlandic resources. Instead it is based on the social and economic intervention of the Danish state. First of all the plans G-50 and G-60 (reports of the Greenland Committee from 1950 and 1960). And these plans were mainly developed in a Danish context by Danish politicians, civil servants and, which is important, based on Danish interests, resources and traditions.

From 1979 the Home Rule Government started to take over, gradually, the authority of a, in many ways, high developed society from the Danish state. But it did not have the corresponding possibilities to maintain the administrative

236

tasks that followed. And taking over authority also meant that the Home Rule had to employ staff for the preparings of the transitions.

It is obvious that the Greenlandic 'state' has different conditions from what is present in Denmark. Looking at the staff requirement it will be seen that in 1987 the Danish ministerial administration employed less than 8 per cent of the total number of engaged labour. As regards Greenland, the Home Rule Administration employed 25 per cent in the same year.

Therefore the administration was and still is very dependent on the size and abilities of the educational system which is still quite insufficient. G- 50 and G- 60 did not contain plans for an educational system at all levels.
This means for example, that before 1986 it was only possible to take an upper secondary education in Nuuk and Aassiat which is necessary if you want to go to the university. After that three towns got a high school. And not until 1987 it was possible to get an administrative academic degree at the university in Nuuk. In other words most of the Greenlanders who want a higher education must go to Denmark for a longer period.

The administrative tools that were transferred were totally organized units based on a specific historical, ministerial development in Denmark. So it is said in the report behind the Law of Home Rule in 1978:

"...the structure in the present administration is the starting point, because it is considered important that it must later be the task of the Home Rule to change the structure after its own wishes. By this it is achieved that there will be no greater upheavels in the transitional period which could have some practical consequences as regards the continuity in the casework, the conditions of the personnel and the public services."

When I made my investigations only a slight change in the administration could be seen. It still has certain rules, traditions and standards for allocating work and competence/authority that could be compared with the Danish ministerial administration.

This means, among other things, that it demands specific Danish qualifications, often an academic degree, to be placed at high official levels. This in spite of the fact that the Greenlandic clerical staff can have many years of experience in addition to the qualification that they have Greenlandic as their native language.

These demands are not only a part of the formal bureaucratic system in the tranferred administrative units. They are also a part of - and are being supported through - the collective bargainings/agreements between the unions (both Greenlandic and Danish) and the employers (put together in a board - the Home Rule Government, Kanukoka; the association of the Greenlandic municipalities and the Danish state - DOA).

Because of the presence of the unions of the Danish civil servants and their great influence most of the agreements follow Danish standards as regards wages (these are higher than what the Greenlandic unions can achieve) and career/senority system which are, among other things, structured in relation to an educational degree.

There you have a conflict. Besides the one between Greenlanders and Danes, there also is a conflict between the Greenlanders, because also Greenlandic civil servant on high levels are members of Danish unions. This means that you can expect a tendency among these to wish to follow the standard from a Danish context - with its educational graduation.

Where you might find a way to get less independent of Danish labour is for examples in the political decisions that have a specific Greenlandic background. This could be seen in the directorate dealing with the outer districts and small local communities (it does not exist any longer). Its staff policy was to create jobs and working methods which made it possible for as many Greenlanders as possible without an academic degree to be engaged.

Besides a possible conflict because of different wages, another kind of conflict is seen. Most of the Danish employees only work in Greenland for a few years (2-4) often giving the reason that they do not want to be left behind in their career-making by getting only Greenlandic work experience. This means that there are often replacements in the administration at high official level among the leading staff which also affects the stability among the clerical staff.

Talking of a suitable organizational culture, the differences in the wages and the lack of stability does not help to develop such a culture, but these conflicts might on a long view change some of the original Danish demands on the staff.

References

Skau, Mette Marie, 1990, Rammer og betingelser for at mindske af-hængigheden af tilkaldt arbejdskraft i den grønlandske hjemmestyreforvaltning. Speciale. Aalborg Universitetscenter.

Geopolitical Developments and their Significance for Greenland and the Circumpolar Arctic Area

Anne Morits Andersen & Jørgen Taagholt

World War II

It was not until World War II that the Arctic area assumed significance in terms of international security policy. In the colonial era the situation in Greenland was very simple: until as late as 1940 Denmark upheld its very strict isolationist policy, so that Greenland and its people were in practice protected from contact with the outside world. With its naturally isolated remote northern location, the country was well outside the Great Powers' sphere of interest. For Denmark, the primary task of security policy was to ensure Danish sovereignty over Greenland after a dispute with Norway at the International Court in The Hague which ended in 1933 with the granting of full sovereignty over all of Greenland to Denmark. The great Danish scientific activity in Greenland was a crucial factor in the decision.

During the war, the Arctic area, and especially the North Atlantic and White Sea, was drawn into the theatre of military operations thanks to technological developments, and suddenly Greenland had a strategic position. Greenland took on independent importance as a "stepping stone", since at this time no aircrafts were available that could cross the Atlantic without stopovers. The Americans therefore established several air bases in Greenland. Greenland became important, too, for naval engagements in the Atlantic, since it was a source of meteorological and communications data.

Post-war development

The World War was succeeded by the Cold War which resulted in the build-up of the strategic US Air Force. Since the shortest route from North America to the industrial centres of the Soviet Union is over the Arctic area, instead of a winding-down of the base activities in Greenland after the war, there was an expansion, including the establishment of the Thule Air Base in 1952 and of large strategic early warning systems: the DEW line, which runs from Alaska over Canada with four stations (DYE 1-4) in Greenland from Holsteinsborg to Ammassalik (two of them on the ice sheet) then on to Iceland; and the BMEWS (Ballistic Missile Early Warning System) with stations in Alaska, Thule AB and Britain which were later supplemented with other early warning stations.

With the establishment of the Thule Base it became possible for the new American B-47 medium-range bombers to reach important industrial centres in the former Soviet Union with just one refuelling. Moreover, Thule AB was an ideal starting-point for reconnaissance flights over the Soviet areas. Finally, the big build-up of Soviet nuclear weapons and weapons delivery systems had made Britain an unreliable stopover point. With the establishment of the Thule Base, the Western World showed that it was technically possible, even in the High Arctic area, to build up a modern industrial community, and the experience of Thule took on importance for work in Alaska and Arctic Canada.

Thule AB was to play an important role in the strategic balance between the USA and the Soviet Union in the first phase of the Cold War.

The Soviet Union

But the greatest changes took place in the Soviet Union, where, during and after the Second World War, there was a military-industrial expansion unparalleled in Arctic areas. The greatest geographical concentration of forces can be seen in the Kola Peninsula, to which before its collapse the Soviet Union (and presumably Russia subsequently) attached extraordinarily great military and industrial importance. There are presumably still about

500 military aircraft and two mechanized infantry divisions totalling 25,000 men stationed there, spread over twenty military installations. Their tasks are among other things to defend the important civilian and military industrial activities there, and to protect the Soviet Northern Fleet which had expanded throughout the seventies to become the biggest of the Soviets' four armed fleets. The Northern Fleet consists of sixty major surface vessels and about 175 submarines with a total manning of over 100,000. This includes about 70% of the Soviet strategic submarine fleet which can operate out of the Murmansk area under cover of the Arctic ice.

Social changes in Greenland

The whole of this development in military technology and the related developments in security policy led to important social transformations in the circumpolar area.

In Greenland a local wish grew up to open up the country after the isolationist policy of the colonial era, and to allow Greenland to benefit from the technological developments which the world war had given the Greenlanders the chance to observe at close quarters as spectators.
The social changes in Greenland and in many other Arctic areas have been enormous in the areas of housing, employment and communications. Remote, sparsely-populated, isolated communities which only had contact with the outside world a few times a year now have regular flight connections, and telephone and fax links have been established with all settlements.

The social consequences of these strategic and security developments have been great, but there is still little clear understanding in the local communities of the relationships between the large geopolitical situation and local development.

The efforts of the Danish armed forces in Greenland have been very limited in military terms, but the Danish military still carries out a number of tasks which is of great importance to the local community - for example the

243

inspection of fisheries, the rescue service and emergency work in crisis situations. And indeed the attitude to the military in Greenland has been positive, and voluntary military service in Greenland has for many years been a popular form of basic training which in many cases has ensured the people involved permanent employment later. But in some periods (especially the seventies) the US military presence in Greenland has been the object of debate and some antagonism; the outcome of this in the eighties, however, has been general acceptance of the fact that the Danish-American defence agreement also had the aim of safeguarding freedom and democracy for the Inuit of the Western World.

In the course of the 1980s, especially during the Reagan administration, the Americans became more interested in exploiting their own strategic strengths and countering others, and this increased the need for a highly developed communications and early warning system. In addition came the US Strategic Defense Initiative (SDI), popularly called "Star Wars". So a modernization of the Thule radar system was required which was clearly against Soviet interests and meant that the issue was brought up at the arms control talks in Geneva. Nevertheless the modernization was carried out, and in 1987 it was followed up by a joint Danish-Greenlandic declaration on the purely defensive nature of the Thule Base.

The collapse of the Soviet Union

The gradual détente between the USA and the Soviet Union (later Russia) reduced the strategic importance of the Greenland area. Although the radar system at Thule still exists, the incipient collapse of the Soviet Union and the American need for restraint with defence spending led to a weakening of American interest in the installations in Greenland.

After the 1st October 1992, the Thule Base is the only remaining US military installation in Greenland (cf. the Danish/US agreement signed in November 1990).

New geopolitical developments

These latest geopolitical developments which mean that military installations and bases are now being closed down, have crucial practical and operational consequences for many local communities which to some extent have made themselves dependent on the military installations in terms of both employment and the level of services.

These developments demonstrate clearly that geopolitical factors were in many respects determinants of development in Greenland. Particularly important are negotiations on aspects of maritime law which have extended national sovereignty obligations are over huge sea areas with mineral and living resources and created new boundaries with other nations. This has made demands on Denmark's will and ability to safeguard the rights and obligations which are a precondition for preserving the country's international status as a nation.

To this we must add increased national responsibility for the formulation of those international conventions which aim to protect the environment and the balance of nature; conventions which could very easily come into conflict with the industrial exploitation of existing resources.

The extension of sovereignty rights in fisheries and economic zones - and the establishment of boundaries at sea - causes disputes. (Examples are the North Atlantic island of Jan Mayen, St. Pierre and Miquelon on the Newfoundland Banks and the grey zone east of Svalbard. Japan claims rights to the southernmost islands of the Kuril group).

In particular, the wishes of the coastal states - in the longer term - to implement anti-pollution regulations, which may lead to limitations on the traditional free shipping and air traffic outside territorial waters, seem to involve conflictful issues.

The potential disputes could concern economic, environmental or ethnic and religious issues.

The economic factors

The stagnation which is characteristic of the economic development of the OECD countries is also influencing the circumpolar countries. The changes in geopolitical conditions are giving the great multinational industrial concerns new operational areas in Eastern Europe, including Russia; areas where resources have been mapped, where the necessary technology is available, and where Western technology is desirable and necessary. But chaos or instability may be a limiting factor. There is therefore only a limited interest in investing in new Arctic prospecting activities and the development of new, cost-intensive technology - for example Arctic offshore and subsea technology. The sudden braking of the hectic development in many Arctic societies is producing serious economic problems and causing increasing internal political tension as the local communities are given more self-determination or even home rule as in Greenland.

Declining military interest

International détente has attenuated the military importance of the Arctic area, and military installations, air bases, radar early warning systems, etc. are being closed down accordingly. This means that important elements in the operation of the infrastructure in many Arctic societies have suddenly been removed, often with disastrous effects on the local communities which in terms of communications (telecommunications and transport) are being pushed back decades. As for employment, service contract work for the armed forces has also hitherto meant many jobs in the local communities which is presumably why Greenland is now taking an extremely positive attitude to the preservation of the Thule Base.

Economic zones

Technological development has meant that large sea areas which were once beyond the sphere of interest of the local communities are now being drawn in as important fisheries zones or economic zones. Many local communities including the Greenlandic ones have engaged in industrial fishing, so they have their own interest in establishing fishing zones; but this also gives rise to national disputes of the kind we see today between Denmark/Greenland and Norway over Jan Mayen. While it was formerly crucially necessary for allies to cooperate - for example within NATO - the same military incentive does not exist today. So in the coming years we can expect to see an increasing number of disputes, for example between Canada and France over the economic zones off St. Pierre et Miquelon, where France lays claim to large parts of the Continental Shelf south west of Newfoundland.

Environmental factors

The opening-up of Eastern Europe has given us new insight into Eastern European conditions, including the disastrous environmental conditions there which are far worse than the West had imagined. We are today in a situation where millions of people live in areas that must be described as uninhabitable, and new accidents at nuclear power plants or with military or industrial activities may mean that the world will see millions of environmental refugees. For humanitarian reasons and in our own interests, Western Europe must invest in the limitation of the spread of pollution in the East.

The dumping of hazardous material in the Arctic Ocean, for example off the Siberian Islands, is a serious threat to the circumpolar Arctic communities. Recent studies of the air pollution in the Arctic give grounds for concern. Hitherto we have regarded Greenland, for example, as a relatively clean reference area where the environmental history of the Northern Hemisphere for the last 100,000 years is deposited in the Greenland ice sheet. Studies of developments in recent decades show that the air pollution from Central Europe is spreading via Eastern Europe over the Arctic Ocean to Arctic regions.

Arctic communities are therefore increasingly interested in helping to find models for combating a pollution which is no respecter of national boundaries. On the international scale in recent years, several initiatives have been taken to limit pollution.

Animal protection

At the same time, increasing global environmental commitment has incalculable consequences for the Arctic local communities. Thanks to the unchecked industrialized whaling of the industrial countries, the whaling of the globally very modest numbers of indigenous hunting populations has been hit by unreasonable limitations. The campaigns of nature conservation associations have meant that not only truly endangered species, but for example seals too, have been perceived as species at risk. While for millennia the Arctic communities have based their existence on maximum exploitation of seals for food, light, heat and clothing, sealskins are now rotting in huge stockpiles - sealskins that would formerly have made an essential contribution to the financial economy of the hunting society. Similarly, conservation rules and hunting regulations have meant that North American Inuit communities can only kill reindeer for their own use, not for sale, in certain areas, with the result that the number of reindeer is growing and areas are being overgrazed and despoiled.

Developments like these increase the tension between the local communities on the one hand and national governments and international organizations on the other.

Ethnic and religious factors

History affords us innumerable examples of regional conflicts caused by ethnic and religious factors. While the situation in Greenland today is not leading to serious disputes, since the Greenlandic population is relatively homogeneous, there is great potential for conflicts in Russia, Alaska and Canada. The indigenous populations in these areas are in the minority, and the collapse of the Soviet Union has given minority groups the chance to have their say. In Alaska, Amerindians and Inuit have won certain rights and economic compensation for American exploitation of the resources of their territories, and in Canada there is an ongoing debate on greater self-determination for indigenous groups. For historical reasons, Southern Canada is split up into an originally English Atlantic province, then a French-dominated Quebec province, and thereafter, from the Ottawa River, an English-American-oriented West Canada. The tension of recent years between Quebec and the rest of Canada led, after long-lasting, difficult negotiations, to a compromise proposal meant to ensure the individual areas, including Quebec, greater independence. However, the referendum of the 26th October 1992 rejected the proposal, since Quebec did not consider the proposal far-reaching enough, while the other provinces thought it went too far. The situation today is so tense that the Canadian state could easily break up, with all the unforeseeable consequences that would involve.

Conclusion

While strategic developments in the post-war years placed the Arctic at the centre of events, the latest geopolitical developments have quite clearly reduced the strategic importance of the Arctic area; but developments in military technology combined with the latest geopolitical developments have provided new food for thought about the Arctic as a traffic node. There is for example international cooperation on a feasibility study of the potential for commercial shipping via the Arctic Ocean. Such a technical and economic development, combined with possible increases in oil and natural gas activity in Arctic areas, would affect security policy development, especially if the Arctic area, contrary to expectations today, is to play a crucial role in global oil and gas supplies.

The dissolution of the Soviet empire has apparently resulted in a very insecure position and an uncertain future for the Russian society. On the face of it, it appears that there is as much of a military presence as before - at least in terms of forces - in the Arctic area, primarily in the Kola Peninsula. However, in the shorter term there is nothing to suggest that in this respect Russia would be able to turn the situation to its own advantage and thus alter the reduced strategic interest in the Arctic area, primarily Greenland.

It is considered to be a fact that in spite of positive results of the disarmament negotiations and destruction of missile systems, great amounts of military hardware are still stored in Russia. In particular, the Kola Peninsula is presumed to continue as a military centre. The chaotic economic conditions in Russia, the insecure political situation and the continuing presence of former, disillusioned or active military personnel create a security and political risk which the western countries should not neglect. The increasing cooperation between the West and Russia, even in the framework of NATO, promises a more positive development. But if the economic situation in Russia becomes more critical, one should remember that history shows that one of the means by which the people are persuaded to accept "unacceptable" conditions is to have powers coming from outside, threatening the nation.

The western world, including the Arctic area, is therefore in a position where it is necessary to create a closer cooperation, support the local development of a sustainable economy, assist with the solution of threatening environmental problems and thus help Russia to participate in an increasingly stable economy and democratic development. At the same time, a reasonable balance must be kept, and a continuing strong NATO should be ready to meet a suddenly changed situation in Russia, should it occur.

In the years to come, security policy in the Arctic area will be determined less by the military superpowers than by economic, environmental, ethnic and religious interests.

References:

Bach & Taagholt, 1982, *Greenland and the arctic region - resources and security policy*, The Information and Wellfare Service of the Danish Defence, København.

Holst, Johan Jørgen, 1991, *Forsvars- og sikkerhedspolitiske udfordringer i den nye verden*, Oslo.

Jane's Defence Weekly, 3rd of March 1990.

Lindsey, George, 1989, *Strategic stability in the Arctic*. Adelphi Papers.

Petersen, Nikolaj, 1992, *Grønland i global sikkerhedspolitik*, Det sikkerheds- og Nedrustningspolitiske Udvalg, København.

Shea, Jamie, 1993, Overview of political changes in the international scene, *AGARD Conference Proceedings*, Technical Information Panel, oktober 1992.

Taagholt, 1986, Greenland and the Faroes. Published in *Northern Waters: Resources and Security Issues* , p.p. 174-189, edited by Clive Archer and David Schrivener, Kent.

Working for Arctic Wilderness

Peter Prokosch

Information about the Arctic Programme of the World Wide Fund for Nature (WWF)

By establlishing an Arctic coordination bureau in Oslo, WWF-International launched its new Arctic Programme in October 1992. Conservation activities round the North Pole of the national WWF organisations in Canada, USA, Norway, Sweden, Finland and Denmark as well as projects in Russia will be reinforced in order to build a strong circumpolar lobby for the Arctic as a whole. Collaboration is also planned with other non-governmental organizations (NGOs) working toward similiar goals. With the recent political developments in Russia and governmental cooperation since 1991 on Arctic conservation under the so-called "Rovaniemi Process", we face the unique chance of creating and supporting panarctic conservation thinking and action.

Political Development and WWF Activities

As the largest shareholder of Arctic environment, the political opening of the former Soviet Union in the late 1980s renewed interests in this polar region and new circumpolar cooperative activities were possible for the first time. Many new Arctic institutions and international joint ventures have been started. Economically, for example, regions in the north have established their own networks as evidenced by the "Northern Forum."

At the same time, on the initiative of Finland, the first panarctic attempt to develop a common Arctic Environmental Protection Strategy was started. With the ministerial conference of all the Arctic countries in Rovaniemi, Finland, in June 1991 the so-called "Rovaniemi Process" was launched.

WWF, with national organizations in most of the Arctic countries (USA, CDN, SF, S, N, DK) and projects in Russia, is predestined to play an essential role in building a panarctic non-governmental lobby for Arctic conservation.

Recent WWF activities in the Arctic countries that could be further developed, reinforced or used as a base for broader collaboration include the following:

Russia

Since 1989, at their first opportunity, WWF-Germany and WWF-International have been cooperating with Russian authorities, scientific institutions and NGOs at the central, regional and local levels to support nature conservation projects in northern Siberia. Most efforts have focused on establishing a new 50,000 km^2 nature reserve ("sapovednik") called the Great Arctic Reserve on the northern coast of the Taymyr peninsula.

As Taymyr is the largest Eurasian tundra region, another key working area was chosen in 1991/92; it is the Lena River delta and is considered the most prominent Arctic river delta in Eurasia. WWF-D/Int supported the existing Lena delta sapovednik (28.000 km^2) with equipment and scientific and political support; there are also plans for an enlarged biosphere reserve which will include the New Siberian Islands.

Activities in northern Siberia (WWF-Int Project 4649/RU0004) have inspired various governmental and non-governmental involvement as well as the scientific interest of several nations, in particular the Netherlands and Germany due to the relationship with to the Wadden Sea/East Atlantic Flyway of coastal birds but also the UK, Norway, Finland, Sweden and France.

Canada

In North America WWF-Canada, especially with their Endangered Spaces Campaign, has been most active in conservation projects in the Canadian Arctic archipelago. WWF-Canada's goal is to complete the representative protected areas network across Canada by the year 2000. The latest national park agreement in the Arctic, called "Aulavik," occupies 12,000 km^2 on Banks Island. Special efforts are in place to employ conservation activities designed to work in partnership with the Inuit peoples. The First Nations have gained substantial power and resources and are involved in a wide variety of co-management agreements and agencies regarding renewable resources.

USA/Alaska

In southern Alaska, WWF-US is paying attention to the endangerment of the Kodiak bear in the Kodiak National Wildlife Refuge. They are operating a project to buy protect land that would otherwise be sold for development.

Furthermore, USA and the former Soviet Union have been cooperating for some years on a challenging plan to design an "international park" across the Bering Sea. WWF-US has been involved in the efforts to establish this regional network of existing and planned reserves through its support of the Alaskan office of the Audubon Society. However, information from Russia shows little positive development in this project.

Of further concern is that living resources in the Bering Sea are at present being heavily exploited by fishing fleets from several nations. Several fish stocks are declining rapidly. Populations of various other groups such as sea lions, fur seals, harbor seals, walrus and various seabirds are also in decline. A new approach for a special WWF Beringa engagement is highly recommended.

Denmark/Greenland

WWF-Denmark has been most active in Greenland, with its 900,000 km^2 national park the world's largest nature reserve. The park today could be used as a reference for other Arctic protected areas. Recent projects have dealt with saving whales, assessing and monitoring seabird populations, and the protection of Falco rusticolus. There are also intentions to open a WWF regional office in Greenland.

Finland

In the tradition of the WWF Mission, WWF-Finland's efforts focus on the conservation of biological diversity, sustainable use of natural resources, and the prevention of wasteful consumption and pollution. The specific problems they are addressing in the north include intensive forestry, building up of rivers, excessive reindeer grazing, poaching, tourism and transboundary air pollution from the Kola Peninsula.

Norway/Svalbard

WWF-Norway's relevant Arctic area is the Barents Sea region. Apart from a tourism project at Ny Lesund/Svalbard and plans for further involvement in issues of eco-tourism in the Svalbard region, WWF-Norway supports the work of "Kontaktutvalget For Nordomrdene," a highly efficient joint umbrella group coordinating the work of several Norwegian NGOs. Together with Russian authorities and NGOs, they are working toward creating an protected international wilderness area in the Barents Sea including Franz Josef Land, Svalbard and larger parts of the marine shelf.

Sweden

WWF-Sweden have been very active in political and lobbying efforts in support of the fundamental Rovaniemi process. They took the initiative for the first WWF Planning and Strategy meeting held in WWF-Denmark's offices in Copenhagen on 11 October, 1991. The representatives from WWF-Denmark, Germany, Norway, Sweden and the US who participated in this meeting felt that the time had come to coordinate the various WWF activities in a direction that a strong panarctic conservation lobby would result. They recommended that WWF-International should establish an Arctic Programme that would coordinate international policy work as well as various field activities, and that a WWF Arctic Subregional Team (AST) with representatives from WWF-International and the WWF NOs to guide the implementation of the project.

International

This proposal was approved by WWF International. The Arctic Programme has a mandate in effect until August, 1995. WWF-Norway is hosting the bureau and WWF-Germany has offered significant financial support and even lent one of its staff members to WWF-International to coordinate the programme for the next three years.

The value of Arctic nature and threats that it faces

Not only do the polar regions and their physical conditions influence the global influence the global climate significantly, and therefore its ecological state, they also present the world's largests wild nature areas. Referring to terrestrial vegetated ground which is not permanently covered with snow or ice, there is no wider wilderness zone remaining on earth than the Arctic Tundra together with the adjacent Taiga. The Arctic seas belong to the group of the most productive ones on earth. The high rate of fish harvesting in the Barents, Bering and Greenland Seas expresses the value for fish resources here and in adjacent waters. Impressive cliff colonies of millions of fishing sea birds (of these Plautus alle may be the world's most common bird species) and huge populations of Arctic marine mammals such as walrus, seals and whales are another picture of the richness of the Arctic. With a population of 20,000 - 40,000 polar bears present a well known and once endangered peak of the food chains in the Arctic Ocean.

There is hardly another area where it is better shown how water shapes a terrestrial landscape than in the Arctic tundra. With its unique polygon structures, or permanently changing meandering streams, or erroding runoffs the tundra still serves as the elementary example of how nature forms itself on a large scale without physical human interference. Biological phenomena characteristic of the tundra are the huge herds of migrating reindeer and caribou and their population interrelationship with the vegetation cover or the cycle population dynamics of lemmings being of vital importance to the reproduction of other mammals and birds. More wetland birds such as divers, geese, swans, ducks and waders breed in the tundra than anywhere else; for example, all Calidris species which show up in high numbers of individuals at other seasons of the year in most wetlands of the world. The main flyways of coastal birds stretching along the continents from the northern to the southern hemisphere start from the breeding grounds in the circumpolar tundra.

The long survival of the wild nature in the Arctic seems to be due to the fact that life in this human-hostile environment is difficult and often not economic. Modern technology and transportation may change this by bringing more people to the north and making more use of its natural resources which may also serve to endanger the social systems and the subsistence life of the indegenious people.

Apart from global threats due to micropollutants and changes in the atmosphere, several concrete and potential threats to the Arctic can be seen:

• Physical destruction of wilderness landscapes due to mining activities, large scale development of oil and gas exploration and extraction, transport roads and other lines (e.g. oil exploration plans for the Arctic National Wildlife Refuge/Alaska, plans for new coal mines and roads south of Longyearbyen/Svalbard). Thereby it has to be noted that developments in Siberia in the past have in many cases been more irresponsible than in "western" countries, as they have often destroyed and polluted the natural landscape while wasting resources at the same time. The introduction of environmentally sophisticated technology to already active industrial sites therefore has to be seen in a different light than developments in any, so far, untouched areas.

• Military activities and nuclear waste (note in particular large scale nuclear dumping in the eastern Barents Sea offshore Novaya Zemlya).

• Unpredictable developments in connection with the international opening of the Northern Sea Route.

• Exploitation of living sea resources such as fish and whales (note, for example, that the populations of Bowheads and Atlantic Right Wales have yet to recover from the old whaling times and the impact that overfishing in the 1980s in the Norwegian/Barents Sea had on sea bird colonies).

• Local hunting and fishing: note the decreasing and endangered goose populations in western Alaska and eastern Siberia; increasing industrial fishing interests for the rich (partly endemic) fish fauna in the Lena delta; more and more uncontrolled hunting tourism in Siberia; feeding of polar foxes in large fur farms in Chucotka with the meat of grey whale and walrus.

• Increasing domestication of reindeer and associated problems of over-grazing as observed in Finmark; also new domestic stocks in northern Siberia get in contact with the wild herds.

• Selling off of wildlife habitat as is happening within the Kodiak Refuge of Alaska thus endangering the Kodiak Bear.

• Increasingly uncontrolled developments in Arctic tourism and the use of questionable and harmful means of transportation, for example there are Russian nuclear icebreaker tours to every destination around and including the North Pole; snowscooters and track vehicles also leave traces on the fragile tundra surface for decades.

Tasks of WWF's Arctic Programme

General tasks and methods: As mandated to create a common panarctic thinking, as well as acting on and supporting conservation steps and developments in this wild region, WWF urgently needs to evaluate all possibilities to originate an easy understandable public image of what the Arctic means. Circumpolar assessments of natural values, as well as classified threats, conservation projects and solution models for environmental problems have to be visualized and mapped.

As there are hardly surveyable local, national and regional Arctic conservation activities ongoing, the main task of WWF being an international organization present in nearly all the Artic nations will be to build up a circumpolar lobby for the Arctic nature as a whole. Therefore not only more intensive cooperation of the different national organisations of WWF is essential, but also collaboration with all other NGOs heading in the same direction. Providing and supporting a circumpolar information exchange on Arctic conservation issues at the same time, while also bringing forth constructive ideas, will be a leading way to achieve this goal. Collectively caring for the most natural biome in the garden of our developed countries will be quite a challenge but should offer a convincing example for similiar efforts in other, warmer regions and developing countries.

As many countries possess competent polar research instititions, and different organizations and governmental bodies care for all types of scientific, social, economic and political issues in the Arctic field, WWF's preference is to identify these resources, interlink with them and support any form of (panarctic) information exchange and cooperation that is relevant and useful for Arctic conservation. In-house investigations concentrate on contributions to a circumpolar protected area network and solutions to sustainable living or have a pioneering and catalytic nature, in order to achieve greater, governmental nature conservation engagements and obligations.

Priorities: To implement the WWF Mission and its philosophy of "Caring for the Earth" in the region will offer the general background. Thereby the following gradations of priority serve as guidelines in the Arctic:

1. Preserving wilderness (nature for its own sake) will be given highest priority. The large remaining coherent untouched Arctic wilderness areas should be defended against any intentions to cut it into pieces or to devaluate it by physical interference. Where possible the Antarctic conservation strategy should also be applied to the Arctic.

2. *Sustainable living* will be proposed as a method to optimize human living demands while preserving natural resources. Particularly in areas where indegenious people still demonstrate a life in harmony with nature, much attention should be given to save these social structures. Sustainability must be the basic principle for any fishing and hunting activity in the whole Arctic.

3. *Restoration and cleaning of destroyed or polluted industrial areas* will be an issue about which WWF wants to increase awarness, particularly in Russia, in order to stimulate governments and industries to invest more effort in techniques that are safer for the environment. There are quite a few cases where sophisticated techniques could promote a wiser use of resources while bringing the natural conditions of a landscape back to higher qualities.

Finally,
Special tasks:
a) Promoting the "Rovaniemi Process" in order to

● formalize international cooperation on and committment to the protections of the Arctic environment through the negotiation of a legally binding Convention on the Protection of the Arctic Environment;

● develop a specific protocol for the Convention on nature conservation and biodiversity in the Arctic region;

● establish a network of representative protected areas (terrestrial and marine) in the Arctic region;

● include the management of living marine resources in the international cooperation;

● establish a permanent international secretariat for the implementation of the strategy and the convention with its protocol.

261

The international coordinating function of WWF should support the commonly formulated statements of NGOs and promote them as equally important as official contributions of the inter-governmental process. Obtaining official observer status in the Rovaneimi Process for the NGOs will be important in that context.

b) Production of various lobbying materials
In order to build a panarctic conservation lobby various types of material are forseen:

• a *booklet* on the Arctic highlighting the unique characterictics of Arctic ecosystems, emphasizing the emerging threats against the Arctic environment and describing WWF activities in field projects in the Arctic region so far;

• an *Arctic conservation newsletter* with concepts similiar to WWF's well established Wadden Sea and Baltic newsletters;

• *posters, slide shows, films* and other informational material;

• a *comprehensive study report* on the common future of the Arctic which will include and summarize the most up-to-date information by leading Arctic experts (an analogue process has been realized in Denmark, Germany and Holland, in a report called the "Common Future of the Wadden Sea").

c) Design and implementation of WWF NOs field projects in a common panarctic context
National organizations of WWF are asked to continue contributing to their own Arctic projects and raise funds for them. Special emphasis should be given to international regions like the Bering Sea and Barents Sea and to relate activities to our general Arctic tasks.

WWF NOs of the Arctic countries active in this region should send a representative to the AST who can report, advise and, if possible, function as a national coordinator, thus building cooperation with other NGOs. In this way a circumpolar information network will be achieved.

d) Projects in Russia will be given high priority. If the goals of the current project in northern Siberia (project number RU0004 mentioned earlier) are realized they may serve as good examples for further engagements in key areas of Russian Arctic tundra and coastline.

Appendix A: List of the participants

Jes
Adolphsen

Aalborg
University

Strandvejen 19
Dk-9000 Aalborg
Denmark
Phone : +45 98 13 87 88
Fax : +45 98 16 88 58

Anne Morits
Andersen

Danish
Polar Center

Strandgade 100 H
DK-1401 København K
Denmark
Phone : +45 32 88 01 00
Fax : +45 32 88 01 01

Tønnes O.K.
Berthelsen

Copenhagen
Business School

Løvstræde 6 , P.O.Box 1042
DK-1007 København K
Denmark
Phone : +45 33 91 12 12
Fax : +45 33 15 75 90

Anders
Bilgram

Arctic Forum i/s

Chr. Købkesgade 3
DK-8000 Århus C
Denmark
Phone : +45 86 12 84 69
Fax : +45 86 12 84 69

Finn
Breinholt Larsen

University of
Greenland

Morbærvej 30
Dk-8260 Viby J.
Denmark
Phone : +45 86 13 01 11
Fax : +45 86 19 44 63

Jørgen Ole   North Atlantic   Building 21.2 , P.O.Box 260
Bærenholdt   Regional Studies   DK-4000 Roskilde
                  Roskilde   Denmark
                  University   Phone : +45 46 75 77 81
                                        Fax : +45 46 75 42 40

Jens   IWGIA   Fiolstræde 10
Dahl   DK-1171 København K
                                        Denmark
                                        Phone : +45 33 12 47 24
                                        Fax :

Susanne   Center for North   Finlandsgade 26
Dybbroe   Atlantic Studies   DK-8200 Århus N
                  Århus   Denmark
                  University   Phone : +45 86 16 52 44
                                        Fax : +45 86 10 82 28

Peter   North Atlantic   Building 21.2 , P.O.Box 260
Friis   Regional Studies   DK-4000 Roskilde
                  Roskilde   Denmark
                  University   Phone : +45 46 75 77 11
                                        Fax : +45 46 75 42 40

Niels   Tordenskjoldsgade 38
Grann   DK-4200 Slagelse
                                        Denmark
                                        Phone : +45 53 52 76 02
                                        Fax :

Tom   Aalborg   Fibigerstræde 11
Greiffenberg   University   DK-9220 Aalborg
                                        Denmark
                                        Phone : +45 98 15 85 22
                                        Fax : +45 98 15 65 41

| Hans Gullestrup | Aalborg University | Fibigerstræde 2
DK-9220 Aalborg
Denmark
Phone : +45 98 15 85 22
Fax : +45 98 15 65 41 |

| Sussi Handberg | Aalborg University | Fibigerstræde 2
DK-9220 Aalborg
Denmark
Phone : +45 98 15 85 22
Fax : +45 98 15 32 98 |

| Ole Hertz | | Gudhjemvej 50
DK-3760 Gudhjem
Denmark
Phone : +45 56 48 54 28
Fax : +45 56 48 54 28 |

| Jack Hicks | Inuit Tapirisat of Canada | 170 Laurier Ave.W,Suite 510
Ottawa , Ontario K1P 5V5
Canada
Phone : +1 613 238 8181
Fax : +1 613 234 1991 |

| Grete K. Hovelsrud | Department of Anthroplogy Brandeis University | Brown 228 , Waltham
Massachusetts
02254 - 9110 USA
Phone : +1 617 736 2210
Fax : |

| Keun Hwang | Greenland Statistical Office | P.O.Box 1025
DK-3900 Nuuk
Greenland
Phone : +299 2 30 00
Fax : +299 2 29 54 |

Urban Ignaz Hügin	Department of Geography Basel University	Hohle Gasse 20 CH-4104 Oberwil Switzerland Phone : +41 61 401 2497 Fax : +41 61 401 1712
Olli Pekka Jalonen	Tampere Peace Research Institute	P.O.Box 447 SF-33101 Tampere Finland Phone : +358 31 23 25 35 Fax : +358 31 23 66 20
Ivar Jonsson	Nat. Institute of Soc. and Eco. Research	Vatnsendabletti 70 A IS-101 Reykjavik Iceland Phone : +354 1 67 71 98 Fax : +354 1 62 88 65
Jyrki K. Käkönen	Tampere Peace Research Institute	P.O.Box 447 SF-33101 Tampere Finland Phone : +358 31 23 25 35 Fax : +358 31 23 66 20
Lise Lyck	Copenhagen Business School	Nansensgade 19 , 5. DK-1366 København K Denmark Phone : +45 38 15 25 75 Fax : +45 38 15 25 76
Gunnar Martens	The Prime Ministers Department	Prins Jørgens Gaard 11 DK-1218 København K Denmark Phone : +45 33 92 33 00 Fax :

Poul
Møller

Center for North
Atlantic Studies
Aarhus University

Finlandsgade 26
DK-8200 Århus N
Denmark
Phone : +45 86 16 52 44
Fax : +45 86 10 82 28

Jógvan
Mørkøre

Faroe Islands
University

Nóatún
FR-100 Tórshavn
Faroe Islands
Phone : +298 1 88 91
Fax : +298 1 68 44

Vladimir
Pavlenko

Arctic Research
Institute
Russian Academy

4HS Shvernik
Moscow
117036 Russia
Phone :
Fax :

Jan H.
Pedersen

Copenhagen
Business School

Blågårdsgade 29 E , 5.tv
DK-2200 København N
Denmark
Phone : +45 31 35 11 79
Fax :

Hanne
Petersen

Greenland
Environmental
Research Institute

Tagensvej 135 , 4.sal
DK-2200 København N
Denmark
Phone : +45 35 82 14 15
Fax :

D. Peter
Prokosch

WWF - International

Kristian Augustsgate 7 A
N-0130 Oslo
Norway
Phone : +47 2 20 37 77
Fax : +47 2 20 06 66

Hans-Erik Rasmussen	University of Copenhagen	Violvej 11 DK-2820 Gentofte Denmark Phone : +45 33 91 21 66 Fax :
Rasmus Ole Rasmussen	North Atlantic Regional Stidies Roskilde University	Building 21.2 , P.O.Box 260 DK-4000 Roskilde Denmark Phone : +45 46 75 77 11 Fax : +45 46 75 42 40
Odd R. Rogne	IASC	P.O.Box 158 N-1330 Oslo Airport Norway Phone : +47 2 123 650 Fax : +47 2 122 635
Mette Marie Skau	Aalborg University	Strandvejen 19 DK-9000 Aalborg Denmark Phone : +45 98 13 87 88 Fax :
Finn Sollie	The Northern Perspectives Group Oslo	Ullernveien 38 N-0280 Oslo Norway Phone : +47 2 50 23 22 Fax : +47 2 50 23 22
Birgitte Sonne	University of Copenhagen	Tyvelse Bygade 16 DK-4171 Glumsø Denmark Phone : +45 53 64 68 07 Fax :

Marianne A. Center for Burnside Hall 720
Stenbaek Northern Studies 805 Sherbrooke St. West
  McGill University Montreal , Quebec
  H3A-2K6 Canada
  Phone : +1 514 398 6052
  Fax : +1 514 398 8364

Axel Kjær Institute of Århus Universitet
Sørensen History DK-8000 Århus C
  Århus Denmark
  University Phone : +45 86 13 67 11
  Fax : +45 86 19 16 99

Bo Wagner Greenland National P.O.Box 145
Sørensen Museum and Archives DK-3900 Nuuk
  Greenland
  Phone : +299 2 26 11
  Fax : +299 2 26 22

Monica University of P.O.Box 122
Ternberg Lapland 96101 Rovaniemi
  Finland
  Phone : +358 960 324 677
  Fax : +358 960 324 600

Eivind Östersund P.O.Box 373
Torp University S-83125 Östersund
  Sweden
  Phone : +46 63 15 53 00
  Fax : +46 63 15 54 54

Jørgen Taagholt	Danish Polar Center	Strandgade 100 H DK-1401 København K Denmark Phone : +45 32 88 01 00 Fax : +45 32 88 01 01
Jan Vejle	Arctic Forum i/s	Chr. Købkesgade 3 DK-8000 Århus C Denmark Phone : +45 86 12 84 69 Fax : +45 86 12 84 69
Hans Weltzer	Institute of Psychology , Århus University	Finlandsgade 26 DK-8200 Århus N Denmark Phone : +45 86 17 55 11 Fax : +45 86 17 59 73
Vagn Wåhlin	Center for North Atlantic Studies Århus University	Finlandsgade 26 DK-8200 Århus N Denmark Phone : +45 86 16 52 44 Fax : +45 86 10 82 28

Appendix B: A list of all prsented papers

Is Greenland Capitalistic?
Jes Adolphsen and Tom Greiffenberg

Geopolitical Developments and their Significance for Greenland and the circumpolar Arctic Area
Anne Morits Andersen & Jørgen Taagholt

The Quiet Life of a Revolution - Greenlandic Home Rule 1979-92
Finn Breinholt Larsen

The Hunting of Marine Mammals in the Kangerlussuaq Area in East Greenland
Niels Grann

Specialization and Dependency in the Use of Arctic Biological Ressources
Ole Hertz

The Inuit Opposition to the Nunavut Land Claim and **Capitalism, Colonialism and Development Planning**
Jack Hicks

On the Concept of Mixed Cash/Subsistence Economy among Arctic Hunters
Grete K. Hovelsrud

Sustainable Development in the Arctic: Inherent Problems and Possible Solutions in a Theoretical Perspective
Urban Ignaz Hügin

Security in the Arctic: Project Conclusions
Olli Pekka Jalonen

Collective Entrepreneurship and Microeconomies
Ivar Jonsson

Political Development in the Arctic; from State to Civil Society
Jyrki K. Käkönen

Considerations on Sustainable Development in the Arctic
Lise Lyck

Unionist and Nationalist strategies in the Faroese EC-debate - Political and Economic Perspectives
Jógvan Mørkøre

Trends in Development and Research in the Arctic Regions of Russia
Vladimir Pavlenko

Environmental Regulation and Control in Relation to Mineral Resources in Greenland
Hanne Petersen

Working for Arctic Wilderness
Peter Prokosch

Cultural Sustainability - Anthropological Perspectives on Terrestrial Animal Production Systems in Greenland
Hans-Erik Rasmussen

Implementation of Sustainable Development
- Methodological and Conceptual Considerations Concerning the Measuring of Sustainability
Rasmus Ole Rasmussen

The Northern Quebec Isolated Communities Database: Structure, Issues, Challenges
Pierre Saint-Laurent

The Conditions for Employment of Greenlanders in the Home Rule
Administration - Especially at High Official Level
Mette Marie Skau

Security in the Arctic - a View from the Scandinavian North
Finn Sollie

Cultural Aspects of Greenlandic Sheep-farming in Retrospect
Birgitte Sonne

Canadian Arctic Science: Institutions and Policy
Marianne A. Stenbaek

Culture as Politics: Expiriences from Greenland
Bo Wagner Sørensen

The Rovaniemi Process and New International Actors
Monica Tennberg

Sami Bill up in Smoke - Discourse Strategies within
Sami Ethnopolitics in Sweden
Eivind Torp

Potential for more Diversified Trade and Industry in Greenland
Jørgen Taagholt

From Agriculture to Modern Fishery Society - the Case of the Faroe
Islands
Vagn Wåhlin